HAWKHURST

H k H

D1346409

C333932691

THE WOOD FOR THE TREES

Also by Richard Fortey

The Hidden Landscape: A Journey into the Geological Past
Life: An Unauthorised Biography. A Natural History of the First Four
Billion Years of Life on Earth
Trilobite!: Eyewitness to Evolution
Fossils: The Key to the Past
The Earth: An Intimate History
Dry Store Room No. 1: The Secret Life of the Natural History Museum
Survivors: The Animals and Plants that Time has Left Behind

RICHARD FORTEY

The Wood for the Trees

The Long View of Nature
from a Small Wood

**WILLIAM
COLLINS**

William Collins
An imprint of HarperCollins*Publishers*
1 London Bridge Street
London SE1 9GF
WilliamCollinsBooks.com

First published in Great Britain by William Collins in 2016

1

Copyright © Richard Fortey 2016

The author asserts the moral right to
be identified as the author of this work

A catalogue record for this book is
available from the British Library

ISBN 978-0-00-810466-5

Printed and bound in Great Britain by
Clays Ltd, St Ives plc

All rights reserved. No part of this publication may be
reproduced, stored in a retrieval system, or transmitted,
in any form or by any means, electronic, mechanical,
photocopying, recording or otherwise, without the
prior permission of the publishers.

This book is sold subject to the condition that it shall not, by
way of trade or otherwise, be lent, re-sold, hired out or otherwise
circulated without the publisher's prior consent in any form of
binding or cover other than that in which it is published and
without a similar condition including this condition being
imposed on the subsequent purchaser.

MIX
Paper from
responsible sources

FSC **FSC™ C007454**
www.fsc.org

FSC™ is a non-profit international organisation established to promote
the responsible management of the world's forests. Products carrying the
FSC label are independently certified to assure consumers that they come
from forests that are managed to meet the social, economic and
ecological needs of present and future generations,
and other controlled sources.

Find out more about HarperCollins and the environment at
www.harpercollins.co.uk/green

For Eileen and Stuart Skeates

Contents

Illustrations

Colour plates

The interior of the wood in March. © Birgir Bohm
The same view in April. © Birgir Bohm
A rare white bluebell. © Jackie Fortey
The lesser celandine. © Jackie Fortey
A beech seedling sprouts from the woodland floor. © Jackie Fortey
Inconspicuous flowers of the holly tree in spring. © Rob Francis
Wild cherry in its vernal glory. © Jackie Fortey
Wild cherry flowers. © Jackie Fortey
Lonny van Ryswyck's experiments with natural materials in the
    wood. Lonny van Ryswyck and Nadine Sterk, Atelier NL,
    Eindhoven, the Netherlands. Photo by Walter Kooken
Yellow bird's nest (*Monotropa*). © Sally-Ann Spence
Painting of ghost orchid by Eleanor Vachell copied from an original
    by Ethel L. Baumgartner. In collections of Amgueddfa Cymru-
    National Museum Wales. NMW 49.29.6241
Pale Tussock Moth. © Andrew Padmore
The writer struggles with an identification handbook in the light of
    the moth trap, guided by Andrew Padmore. © Jackie Fortey
Blood Vein Moth. © Andrew Padmore
Purple Thorn Moth. © Andrew Padmore
Satin Beauty Moth. © Andrew Padmore
Speckled wood butterfly. © Andrew Padmore
Peacock butterfly. © Andrew Padmore

Silver-washed fritillary butterfly. © Andrew Padmore

Comma butterfly. © Andrew Padmore

Red kite. © Rob Francis

Brown long-eared bat. © Claire Andrews

Field vole. © Sally-Ann Spence

Hazel dormouse. © Danny Green

Hoard of late Iron Age coins found inside a hollow flint close to the
    wood. HCR6698 The Henley Hoard, © Ashmolean Museum,
    University of Oxford

Small truffles (*Elaphomyces*). © Jackie Fortey

A selection of snails from the wood. © Jackie Fortey

The cherry-picker lifting the writer up into the canopy of the beech
    trees. © Jackie Fortey

Large black terrestrial spider *Coelotes terrestris*. © P.R. Harvey

Slime mould (*Lycogala epidendrum*). © Jackie Fortey

Crab spider *Diaea dorsata*. © P.R. Harvey

A rare crane fly (*Cercophora*). © Andrew Padmore

The view westwards across the Assendon Valley from Henley Park
    with our wood on the skyline. © Rob Francis

Monument to the Knollys of Greys Court in Rotherfield Greys
    parish church. © Jackie Fortey

The manor house of Greys Court. © Jackie Fortey

The Stapleton family painted by Thomas Beach, 1789. © The
    Holburne Museum, Bath/Bridgeman Images

The pelargonium variety 'Miss Stapleton'. © Jackie Fortey

1708 document recording the granting of rights to grub up the
    'Roots & Runts' of beech trees cleared below our wood. Courtesy
    of the Oxford History Centre (catalogue ref. PAR209/13/9D/4),
    from the collection of the parish of Oxford St Mary the Virgin

Magpie toadstool. © Jackie Fortey

Stinkhorn fungus (*Phallus impudicus*). © Jackie Fortey

Sulphur polypore. © Jackie Fortey

Ink caps growing on a rotten beech stump. © Jackie Fortey

Rhodotus palmatus. © Jackie Fortey

Red slug *Arion rufus*. © Jackie Fortey

Black dor beetle *Anoplotropes*. © Jackie Fortey

An uncommon beetle, *Oedemera femoralis*. © Andrew Padmore

Longhorn beetle *Rutpela maculata*. © Andrew Padmore

Sexton beetle *Nicrophorus humator*. © Andrew Padmore

Alistair Phillips turning a wild cherry-wood bowl from timber
   derived from one of our felled trees. © Jackie Fortey

Small-scale charcoal-burning. © Rebecca Fortey

A flint derived from the chalk at the Fair Mile. © Jackie Fortey

The author on a cherry branch. © Jackie Fortey

Henley from the Wargrave Road by Jan Siberechts (1698). Courtesy
   of the River and Rowing Museum, Henley-on-Thames

Wych elm leaf in autumn. © Rob Francis

Male wych elm flowers in early spring. © Jackie Fortey

The drawers of Philip Koomen's cabinet. Photograph © Rob Francis

The Koomen collection cabinet back in the wood. Photograph ©
   Rob Francis

Shield lichen (*Parmelia*). © Jackie Fortey

Bank haircap moss (*Polytrichastrum*). © Jackie Fortey

Text illustrations

p.xii Location map of Grim's Dyke Wood. © Leo Fortey

p.23 Half-timbered cottage in Assendon. © Rob Francis

p 24 Spring view of Lambridge Wood. © Rob Francis

p.40 Thin section of conglomerate pebble. © Jan Zalasiewicz

p.46 'Fiddlehead' male fern. © Jackie Fortey

p.58 Stacked beech trunks. © Jackie Fortey

p.62 Squirrel damage to a beech tree. © Jackie Fortey

p.65 Sedges. © Nina Krauzewicz

p.86 Aurochs, Les Eyzies, France. Photo © Jackie Fortey

p.90 Misericord, St Mary's church, Beverley, Yorkshire. Photo ©
   David Ross, www.britainexpress.com

p.109 Detail of Richard Davis's map of Oxfordshire, 1797. Courtesy
   Simmons & Sons, Estate Agents, Henley

p.112 Ash and cherry bark. © Jackie Fortey

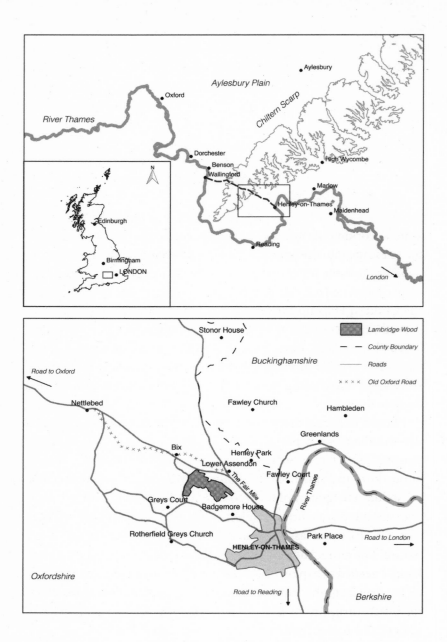

Location of the wood in the Chiltern Hills, with main roads and places.
The old road to Wallingford is marked with 'x's through Bix in the
lower map, dashes in the upper one.

1

April

After a working life spent in a great museum, the time had come for me to escape into the open air. I spent years handling fossils of extinct animals; now, the inner naturalist needed to touch living animals and plants. My wife Jackie discovered the advertisement: a small piece of the Chiltern Hills up for sale. The proceeds from a television series proved exactly enough to purchase four acres of ancient beech-and-bluebell woodland, buried deeply inside a greater stretch of stately trees. The briefest of visits clinched the deal – exploring the wood simply felt like coming home. On 4 July 2011 'Grim's Dyke Wood' became ours.

I began to keep a diary to record wildlife, and the look and *feel* of the woodland as it passed through diverse moods and changing seasons. I sat on one particular stump to make observations, which I wrote down in a small, leather-bound notebook. I was unconsciously compiling a biography of the wood – *bio* in the most exact sense, since animals and plants formed an important part of it. Before long, I saw that the story was as much about human history as it was about nature. For all its ancient lineage, the wood was shaped by human hand. I needed to explore the development of the English countryside, all the way from the Iron Age to the recent exploitation of woodland for beech furniture or tent pegs. I was moved by a compulsion to understand half-forgotten crafts and revive half-remembered words like 'bodger', 'spile' and 'bavin'. Plans were made to fell timber, to follow the journey from tree to furniture; to visit the canopy in a cherry-picker; to explore the archaeology of that ancient feature,

Grim's Dyke, that ran along one side of the plot. I wanted to see if the wood could yield food as well as inspiration.

My scientific soul reawakened as I sought to comprehend the ways that plants and animals collaborate to generate a rich ecology. I had to sample everything: mosses, lichens, grasses, insects, and fungi. I investigated the natural history of beech, oak, ash, yew, and all the other trees. I spent moonlit evenings trapping moths; daytime frolicking with nets to catch crane flies or lifting up rotten logs to understand decay. I poked and prodded and snuffled under brambles. I wanted to turn the appropriate bits of geology into tiles and glass. The wood became a route to understanding how the landscape is forever in a state of transition, for all that we think it unchanging. In short, the wood became a project.

Grim's Dyke Wood is just a segment in the middle of more extensive ancient woodland, Lambridge Wood, lying in the southern part of the county of Oxfordshire. Splitting Lambridge into separate plots generated a profit for the previous owner, but also allowed people of modest means to own and care for their own small piece of living history. Our fellow 'woodies' – as Jackie terms them – proved to include a well-known harpsichordist, a retired professor of business systems, a founder member of Genesis (the band, not the book), a virologist turned plant illustrator, ex-actors turned psychologists, and a woman of mystery. Our own patch is one of the smaller ones. All of the 'woodies' have their own reasons for wanting to be among the trees – some desire simply to dream, some would rather like to turn a profit, others to explore sustainable resources. I believe I am the only naturalist. All the owners are there to prevent the wood from being felled or turned into housing. For the long history of Lambridge Wood tells us that our trees are less worked today than at any time in the past. This sad redundancy is no less part of its tale, as our wood is inevitably connected to the wider world of commerce and markets. The histories of my home town, Henley-on-Thames, a mile away, and the famous river on which it sits, are bound into the narrative of the surrounding countryside. Ancient manors controlled the fate of woodlands for centuries. I have to imagine what the wood would have seen or heard as great events passed it by; who might have lurked under the trees, what poachers and vagabonds, poets or highwaymen.

Once the project was under way a curious thing happened. I wanted to make a collection. This may not sound particularly remarkable, but for somebody who had worked for decades with rank after rank of curated collections it was rejuvenation. Life among the stacks in the Natural History Museum in London had stifled my acquisitiveness, but now something was rekindled. I wanted to collect objects from the wood, not in the systematic way of a scientist, but with something of the random joy of a young boy. Perhaps I wanted to *become* that boy once again. Eighteenth-century gentlemen were wont to have cabinets of curiosities in which they displayed items that might have conversational or antiquarian interest. I wanted to have my own cabinet of curiosity. I would add items when my curiosity was piqued month by month: maybe a stone, a feather, or a dried plant – nothing for the eighteenth-century gent. I believe that curiosity is a most important human instinct. Curiosity is the enemy of certainty, and certainty – particularly conviction that other people are different, or sinful, or irreligious – lies behind much of the conflict and genocide that disfigure human history. If I could issue one injunction to humankind it would be: 'Be curious!' My collection will be a way of encapsulating the whole Grim's Dyke Wood project: my New Curiosity Shop. And I already know that the last item to be curated will be the leather-bound notebook.

The collection requires a cabinet to house it. Jackie and I plan to fell one of our cherry trees and convert it into a wonderful receptacle for the wood's serendipitous treasures. We must discuss the work with Philip Koomen, a noted Chiltern furniture-maker devoted to using local materials. Philip's workshop, Wheelers Barn, is in the remotest part of the Chiltern Hills, only about five miles from Grim's Dyke Wood as the crow flies, but about fifteen as the ancient roads wind hither and thither. His studio is imbued with calm. Polished sections through trees hanging on the walls show the qualities of each variety: colour, texture, grain, and age all combine to distinguish not just different tree species, but individual personalities. No two trees are identical. Some have burrs that section into turbulent swirls. Pale ash contrasts with rich walnut, and cherry with its warm tones is satisfactorily different from oak. This is a man who cares deeply about materials and believes in the *genius loci* – an integration of human and natural history that lends authenticity to a hand-made item. Philip's handiwork from our own cherry tree will be a physical embodiment of our wood, but by housing the idiosyncratic collection picked up as the project develops it will also *contain* the wood, as curated by this writer. It will be a cabinet of memories as much as objects. We haggle a little about design, but I know I shall rely on his judgement. I will have to be patient when I gather up the small things in the wood that take my fancy. It will be some time before the collection can live in its dedicated home.

This book could be thought of as another kind of collection. Extracts from my diary describe the wood through the seasons. I follow H.E. Bates's wonderful book *Through the Woods* (1936) in beginning in the exuberant month of April rather than with the calendar year, and frigid January. But then, H.E. Bates himself inherited the same plan from the writer and illustrator Clare Leighton, whose intimate portrait of her own garden through the cycles of the seasons, *Four Hedges*,[1] he much admired. My friends and colleagues come to sample and identify almost every jot and tittle of natural history that they can find. Natural history leads on to science, and the stories of grand estates, woodland skills and trades, and life along the River Thames. Human folly and natural catastrophes link the wood to a wider world beyond the trees. This complex collection explains why the wood is as it is today; its rich diversity of life is a concatenation of particular circumstances. I am trying to reason how the natural world came to be so varied, and my understanding is refracted through the lens of my own small patch. I am trying to see the wood for the trees.

Bluebells

Some trees stand close together, like a pair of friends huddling in mutual support. Others are almost solitary, rearing away from their fellows in the midst of a clearing. The poet Edward Thomas described 'the uncounted perpendicular straight stems of beech, and yet not all quite perpendicular or quite straight'.[2] Each tree trunk has individuality, for all the harmony of their numbers together. One leans a little towards a weaker neighbour; another carries a scar where a branch fell long ago; this one has an extraordinarily slender elegance as it reaches for its place under the sky; that one has a stocky base, as chunky as an elephant's leg, and doubtless at least as strong. No two trees are really alike, yet their collaboration on the scale of the wider wood creates a sense of architectural design. The relatively pale and smooth beech bark helps to unite the structure, for in the early spring's soft sunshine the tree trunks shine almost silver. The natural cathedral of the wood is supported by brilliant, vertical superstructure, one that shifts subtly with the moods of the sun.

It is too early in the month for many fresh beech leaves to have unfurled from their tight buds. The wood is still flooded with light. Some of the sunlight falls on the crisp, dark tan to gold-coloured leaves fallen from last year's canopy that lie in scruffy patches on the ground; stubbornly dry, they have yet to rot away. The sunshine brings the first direct heat of the year, enough to warm our cheeks with hints of seasons to come. Am I imagining that the beech trunk is actually hot on its illuminated side? It does not strain the imagination to envisage the sap rising beneath the grey-green roughened bark, rejuvenated by April showers. Where the sunlight reaches the thin soil spring flowers accept the warmth and light; briefly, it is their time to flourish. After standing to contemplate the grandeur I now have to get down on my hands and knees to see what is happening at ground level. By the pathside are patches of heart-shaped leaves mottled with white; the sun glances off the tiny, glossy, butter-yellow petals of the lesser celandine, eight petals in a circle per flower, not unlike a child's first drawing of what every flower *should* be. Celandines are growing in the company of dog violets, whose flowers are as complicated as the celandine's are simple: the whole borne on an arched-over stem, carrying five blue-violet petals, of which the lowest is lip-like and marked with the most delicate dark lines converging towards the centre of the flower. At the very heart of it there is a subtle yellowing and, behind, a spur offering a treasure of nectar: clearly the whole structure is an inducement for pollinating insects – a road map promising a reward. Through the beech litter brilliant green blades of a grass, wood melick, are pushing upwards to seek their share of precious light. Near the edge of the wood, lobed leaves of ground elder form a mat of freshest green; this notorious garden weed is constrained to behave itself in the wild.

But close observations of wayside flowers may be something of a distraction from a Chiltern spectacular. Perhaps I prefer to taste a few appetisers before becoming overwhelmed by the main course. For just beyond a short sward of wood melick is a shoreline edging the glory of the April beech woods in England: a sea of bluebells. The whole forest floor beyond is coloured by thousands upon thousands of flowers; a sea – because it seems unbroken and intense, like the yachted waters in a Dufy painting. But the display might equally be described as a carpet of bluebells; that word is more appropriate to the floor of a natural cathedral. Besides, the hue is a dark and rich blue, a shade not truly belonging to the ocean. Rather it is the cobalt blue of decorated tin-glazed wares produced at Delft, in Holland. In these woods, a magical slip is washed over the floor of the woodland as if by the hand of a master; a glaze that lasts only a few weeks, but transforms the ground beneath the beeches. From a distance there is a vague fuzziness about it, as if the blueness were evaporating upwards. The show is produced by massed English bluebells, *Hyacinthoides non-scripta*, a species unique to western Europe. This is old Britain's very own, very particular and extraordinarily beautiful celebration of early spring. There is no physical sign in our wood of the Spanish bluebell interloper *Hyacinthoides hispanica*, the common species in English gardens. It is a coarser plant, with a more upright spike of flowers, and generally less elegant. In many places it is hybridising with the native species.

Each bluebell arises from a white bulb the size of a small tomato, and produces a rosette of spear-like leaves and a single flower spike; none is much taller than your forearm. It takes hundreds to make a splash of colour. The bells hang down in a line along the raceme in a single graceful curve. 'Raceme' is scientifically correct, of course, but how I wish that we could refer to it as a 'chime'. Flowers at the base of the spike open first, their six delicate petals curving backward to form a skirt that curves away from the creamy anthers; it takes a while for the whole display to be over, as each flower up the spike comes to perfection one after another. With a natural variation in flowering times according to aspect and local climate, there are a few woodland nooks where bluebells open up precociously, and others where they linger longer. But wherever they bloom, theirs is a short-lived glory; and only when they are seen in numbers can the delicate perfume they produce be appreciated. As they generally reproduce from a slow multiplication of their bulbs, rather than from seed, the masses of English bluebells seen in our woods are a reliable indicator that they are of ancient origin. Hence it is likely that the flowers that delight my eyes today have been admired for centuries from the same spot near the edge of this very wood. The temptation is to pick a great bunch of blooms, but in a vase they lose vibrancy; they need a myriad companions to assert their natural magnificence. A thrush singing in mellow, repetitive phrases from deep within a holly bush adds some sort of blessing.

This is our own piece of classic English beech woodland, gifted with bluebells, covered with trees for generations, and changing at the slow pace of sap rising and falling. When Lambridge Wood was subdivided by the previous owner our little piece of it was arbitrarily christened Grim's Dyke Wood, after the ancient monument that passes through the wood. The new name added an irresistible whiff of romance to the sales pitch that helped us to part with our money. It is a triangular plot with nearly equal sides, two of them marked out by public footpaths. We access the north-east corner of the triangle by vehicle along a track through Lambridge Wood that leads to a converted barn adjacent to our piece of woodland; the barn has a picturesque cottage next door that will also feature in this book. On the ground, it is hard to detect a very gentle slope of the whole plot to the south, but the incline is enough to admit a magical influx of winter light in the afternoon with the setting sun. About four acres (1.6 hectares) of woodland is not exactly a vast tract of forest, but it is enough to include more than 180 mature beech trees, which I counted, and bluebells galore, which I didn't.

Lambridge Wood sits high in the Chiltern Hills, thirty-five miles to the west of London, near the southern tip of the county of Oxfordshire. Although so near the capital, the wood could be ten times further away from it and would not gain a jot more feeling of remoteness. As I contemplate the bluebells only the occasional growl overhead from aeroplanes bound for Heathrow reminds me that there is a great urban sprawl so close to hand.

The Chiltern Hills form the high ground for a length of more than fifty miles north-west of London. They follow the course of the outcrop of the pure white limestone known as the Chalk.[3] The same rock makes the white cliffs of Dover, where England most closely approaches continental Europe; the sight of the cliffs has brought a lump into the throat of many a returning traveller, so it might be thought of as a peculiarly English rock, although it is actually widely spread around the world. As limestone goes, it is a very soft example of its kind – one that can be flaked with a penknife. Even so, it is harder and more homogeneous than the rocks that underlie it to the north, or overlie it in the direction of London, and differential weathering and erosion over hundreds of millennia has promoted the relative recession of the softer rocks to either side. The Chilterns stand proud.

The scarp slope along the northerly edge of the hills is surprisingly steep, and from the top of the Oxfordshire segment there are fine long views across the Vale of Aylesbury towards Oxford in the distance. That scarp lies only ten miles north-west of our wood. Half that distance away, Windmill Hill at Nettlebed is one of the highest points in southern England at 692 feet (211 metres) above sea level.[4] The tops of the Chiltern Hills are richly wooded compared with the intensively farmed lowlands on the gentle plain to the north, where a patchwork of neat green fields or brown ploughed farmland is the rule. Google Earth or the Ordnance Survey map reveal much the same pattern, whether seen from above or in plan. The high ground has long fostered special pastoral practices, in which woods played a continuous and important part in the rhythm of country life. That is why they have survived. Our tiny patch is just one small piece of a larger tapestry stitched together from irregular swatches of trees, stretching over many miles. Other kinds of farmland are interspersed, to be sure, and in some places there has been sufficient clearance for open downland. But near our patch, copse, shaw, hanger and wood dominate the landscape.

When I first walked through Lambridge Wood as a newcomer to the Chiltern Hills I was overwhelmed by a feeling of entering a realm of eternal nature. Here was the antidote to jaded city life. The woods are unchanging; they help to put our small concerns into perspective. They are restorative, havens for animals and plants; safe places for the spirit. Such a perspective drenches Edward Thomas's rapt accounts of woodland in *The South Country*, and has a modern mirror in Roger Deakin's *Wildwood* nearly a century later. Here is the farmer A.G. Street writing in 1933 after listing more than one disappointment of middle age: 'The majesty of the wood remains unaltered. As I wandered slowly through it, the terrific importance of my trouble seemed to fade away. The peace of the wood and the comfort of the still trees soon iron out the creases in my soul.'[5] Surely some comparable emotion lay behind the enthusiasm with which we purchased Grim's Dyke Wood, our own piece of peace. It was a romantic (or even Romantic) notion, and not wrong in its essentials. But, as Henry David Thoreau remarked of the English poets: 'There is plenty of genial love of Nature but not so much of Nature herself.'[6] The wood has indeed given much pleasure, but much of that delight proves to be an intimate examination of nature close to. And I now know that the history of nature is not only natural history. The wood is not eternal – it is a construct, a human product. It was made by our ancestors, modified repeatedly, nearly obliterated, rescued by industry, forgotten and remembered by turn. The animals and plants rubbed along with history as best they could, mostly unconsidered except as meat, fuel and forage: the natural history was part and parcel of the human history. The result is what we see today. Romantic empathy with 'Nature' is all very well, but it does have to brush up against the hard grit of history, which can soon polish off any coating of wishful thinking.

So this book is both romantic and forensic, if such a combination is possible. My diary records the status of the beech trees and the animals and plants, the play of the light, the passage of the seasons, expeditions and people, and the incomparable pleasures of discovery. I have also taken samples from the wood to the laboratory to dissect under a microscope. I have invited help from experts to identify tiny animals – mostly insects – about which I know little. Add to this excursions into historical literature and archives, and much time spent scrutinising scratchy ancient maps, deeds and sales catalogues to understand how the wood fared under management for profit or pleasure, and its place in the economy of estate, county and country. I have interviewed those who have known the woods during long lives. There will be a little geology, and more than a touch of archaeology.

Several of my previous books have dealt with big themes: the history of life or the geology of the world refracted through a personal lens. This book is the other way round: a tiny morsel of a historic land looked at all ways. The sum of all my observations will lead to an understanding of biodiversity – the variety of animals, plants and fungi that share this small wood. Biodiversity does not just belong to tropical rainforests or coral reefs. Almost every habitat has its own rich assemblage of organisms competing, collaborating and connected. What is found today is the result of climate, habitat, pollution or lack of it, history and husbandry. For me, the poetry of the wood derives from close examination as much as from synthesis and sensibility. But I am aware that description alone does not necessarily lead to understanding. This example from what may be Wordsworth's worst poem ('The Thorn') comes to mind:

And to the left, three yards beyond,
You see a little muddy pond
Of water – never dry
I measured it from side to side:
'Twas four feet long, and three feet wide.

The Darwin connection

Despite this dire topographic warning, I must describe the anatomy of the countryside around the wood, since it is crucial to this history. Grim's Dyke (and Lambridge) Wood lies at the top of a locally high ridge, and immediately to the north of it a steep slope runs away continuously downwards to a rather busy road; that part of the incline below the barn has been cleared of trees, and is now occupied by a well-fenced deer park. The main road is partly a dual carriageway running westwards up a typical Chiltern dry valley and serves to connect the nearby small and historic town of Henley-on-Thames, where I live, with the larger and even more historic town of Wallingford thirteen miles away. Wallingford also lies adjacent to the Thames, but between it and Henley the great river takes an enormous southerly bend by way of big, bustling Reading, as if reluctant to breach the barrier of the Chiltern Hills. This it finally does – and most picturesquely – near the village of Goring, about seven miles from Wallingford, where the Chalk cliffs are steep enough on the eastern side of the floodplain to suggest a gorge. Geographers have more prosaically called it 'the Goring Gap'. Robert Gibbings is the most charming writer on this and other stretches of the Thames.[7] I doubt I can live up to his blend of precise natural history with human observations of all kinds. The journey across country between Henley and Wallingford is very much shorter than the distance along the river, a fact that profoundly influenced the development of medieval Henley and its surroundings, including our small wood. Henley played an important part in the transport of goods and people between London and Oxford, and its story is inextricably bound up with that of the River Thames.

There are other ways to locate our wood within the English countryside. Ancient England is a curiously tessellated collage of different patterns of ownership and responsibility. Parish, village and manor all make different claims. Grim's Dyke Wood lies near the edge of the old ecclesiastical parish of Henley-on-Thames, so its original church, as it were, is the fine, thirteenth-century flint-and-stone edifice of St Mary's in the centre of town two miles away.[8] On the way out of Henley in the direction of Wallingford and Oxford the road is dead straight and splendidly bordered with wide grass verges and avenues of trees. This is the Fair Mile, appropriately named, and the ecclesiastical parish extends out in this direction. At the end of the Fair Mile, a minor road forks off to the right along another valley to Stonor, while the main road continues uphill towards Nettlebed and Wallingford. At this point our wood lies at the top of the slope on the skyline to the left (and south). The fork in the road marks the end of the village of Lower Assendon, and is very near to the wood as the red kite flies, which it frequently does around here. Smoke from Assendon chimneys can be smelled in the wood. Assendon also houses the Golden Ball, the closest pub, which seems nearly as ancient as the hills, and is reached by a steep downhill scramble along a path descending from Lambridge Wood. At the top of the hill, and further away from Lambridge, another ancient village with the briefest possible name – Bix – is arranged around a huge common, and is in a different parish.

But more important to our story than either parish or village is the manor. For most of its recorded history Lambridge Wood, including our piece of it, has been part of the manor of Greys. The manor house, Greys Court, is a remarkable survivor, just a mile away from Grim's Dyke Wood. Both the house and the estate are now managed by the National Trust, and thousands of visitors flock there. These benign crowds of pensioners and picnickers arriving by car make it difficult to imagine the house as a remote backwater, but there was a time when the Chiltern Hills were wild and inaccessible. Criminals could go to ground there; religious dissenters could hide there. Greys Court still commands the least urban aspect in the Home Counties. From the garden lawn the modern road is hardly visible, and the view is dominated by a broad, clear valley dotted with sheep and flanked on either side by dense beech woods. It could still be a landscape through which horses provided the only transport, at a time when London belonged to another world.

Although substantial, Greys Court could hardly be described as a stately home. Part twelfth-century fortified castle, and part Tudor mansion, it remained in private hands from medieval times until 1969. In a brick outhouse an extraordinarily ancient donkey wheel resembling some cock-eyed wooden fairground attraction was used until the twentieth century to lift water from a well excavated deep down into the chalk. It is not difficult to imagine how a place like Greys Court might roll with the blows of history, battening down at times of hardship, fattening up in times of plenty. The extensive estate could provide what was needed: sufficient arable land for wheat and barley, pasture for cattle and sheep, from the beech stands fuel and wild game, and good water from the well. Lambridge Wood lay along the northern edge of the estate. Land nearer the big house was more likely to come under cultivation, so the marginal position of the wood doubtless contributed to its long-term survival. It was always useful just where it was.

The parish church for Greys Court, where the grand names belonging to the big house are interred, is a tiny, flint construction with a low tower, close by the road in Rotherfield Greys, a hamlet that also has the second-nearest pub to our wood, the Maltsters Arms. Church and pub can be reached from the wood on public footpaths leading south-wards and crossing open fields for a little more than a mile. I have never met anyone else on these old rights of way. The paths that run along the River Thames just a couple of miles away are crowded with walkers, but the open Chilterns are still the province of the skylark and the stroller. On a clear spring day, the low hills conceal endless possibilities, all of them joyous. The Maltsters Arms is one of those cosy pubs with exposed oak beams on low ceilings, real fires, no background music, and a landlord who actually seems to like his customers.

The church is next to the pub, as tradition demands. A large chunk of its interior is taken up by a side chapel devoted to the monuments of the masters and mistresses of Greys Court, and principal among these is the exuberant and splendid alabaster and marble tomb of Sir Francis Knollys (d.1596) and Katherine, his wife. Their effigies lie side by side praying in formal splendour, while around the tomb seven sons and seven daughters parade in a pious line. Most touching is a tiny baby who died in infancy, whose effigy lies alongside that of his father. Sir Francis was a noted courtier of Henry VIII and Elizabeth I. I like the thought that on his days away from court Sir Francis may have wandered in our wood for pleasure, or maybe hunted game there. On the floor of the nave a brass to Robert de Grey (d.1388) is altogether more modest, even though the manor and church both bear his name. Clad partly in chain mail, his sword by his side, and gauntlets still frozen in metallic prayer, he seems more a grand cipher than a real person.

In the countryside, for many centuries manors and estates were paramount. Those who owned the estates neighbouring Greys provided its society. These nearby manors suffered the same pestilences and plagues, and shared good years and bad. The lords and gentry knew one another, and paid formal and informal visits. They eventually became what my mother would have termed 'county'. From time to time the estates were home to remarkable historical figures; at others their occupants were quietly obscure. The status of peasantry and servants and artisans changed gradually, but all the estates had to absorb the changes, which continue today. The closest estate to our wood – and Greys manor – was Fawley Court and Henley Park to the north: a pigeon could fly from Grim's Dyke Wood into the Park in a minute. To the east lay Badgemore. The fine house has now vanished, and what remains of it is a golf club. While further still to the north a small and perfectly set stately home remains in its own valley; the Stonor family that lives there boasts more than eight hundred years of occupancy, and one of the longest continuous lineages in Great Britain.

Another map ties the wood more closely with Rotherfield Greys than with Henley-on-Thames. Civil parishes are the basic unit of local government, and frequently do not have the same boundaries as the ancient ecclesiastical parishes. They elect councillors, not priests, and their boundaries were sorted out at the end of the nineteenth century to make a more sensible system of local administration. Our wood lies in the civil parish of Rotherfield Greys, even though it is ecclesiastically Henley; this is appropriate to its other links with the big house. It seems that Lambridge Wood was always on the edge of some map or parish or village, which may be a good place to be to pass unnoticed. And like many other woodlands, our wood was also free from tithes: a 10 per cent levy on the income derived from the land once provided the principal source of income to support the local church. Following an Act of Parliament in 1836 a schedule of tithes was compiled across England, and in the Oxfordshire Record Office a map of 1840[9] portrays Lambridge Wood with considerable accuracy. The accompanying ledger prepared by a clerk in best copperplate script declares it 'exempt'. I occasionally put a pound coin in the box at Rotherfield Greys as a token of expiation.

In 1922 Lambridge Wood was sold off from the Greys estate after a history stretching back to Domesday. We have the map detailing 'Lambridge Farm and 160 acres of woodland' which was sold in Henley Town Hall on 26 July to George Shorland, a rich farmer and entrepreneur who had purchased land all around Henley. The modern era of Lambridge Wood had begun, and the unbroken thread leading back to medieval times had been severed. We will meet some of the subsequent owners later on, but now I am going to take a jump to 1969, when Lambridge Wood passed into the ownership of Sir Thomas Erasmus Barlow, Bt, whose heirs owned it until as recently as 2010. I admit that the name meant nothing to me. Sir Thomas was the third baronet to carry the title, and a distinguished naval commander. In a fairly perfunctory way, I started one of those online searches that have become routine for writers, as they have for almost everybody else. I moved backwards in time as far as I could. The First Baronet, another Thomas, had been Queen Victoria's private physician, a man who died dripping with honours, and no doubt had an outstanding bedside manner. The Second Baronet, Sir Alan Barlow, father of

Thomas Erasmus, was scarcely less distinguished as a civil servant in the grand tradition, serving as Prime Minister Ramsay MacDonald's Principal Private Secretary from 1933 to 1934. But then I discovered something that caused the mouse to freeze in my fist. Alan Barlow had married Nora Darwin. A magical name had somehow found its way into the genealogy of the wood. If one thread had been severed, another had been established. It did not take much more research to establish that Nora was the granddaughter of Charles Darwin. So our wood, the subject of my own modest natural history investigations, had recently belonged to a direct descendant of the greatest natural historian of all time.

I happen to know another direct Darwin descendant who worked with me at the Natural History Museum, the botanist Sarah Darwin. The Darwins are an unusually distinguished clan, and the present generation respects the ramifications of the dynasty. Charles Darwin's grandfather Erasmus sired one of those lineages that seem to have done nothing but good in the world: the third Sir Thomas's middle name must have been a nod in the direction of the grand old progenitor. Sarah knows the current baronet, Sir James Barlow; not least, they are both Ambassadors for the Galapagos Conservation Trust, which seeks to protect the world's most famous natural evolutionary laboratory from further damage through foolish exploitation. In the autumn of 2014 Sarah introduced me to Sir James and his sister Monica. I met them both in our wood, and together we traced a path through Lambridge that they had not done for many years. James remembered his grandmother, who, he said, had been dandled on Darwin's knee. So there I was, talking to somebody whose grandmother might have giggled and snuggled into the breast of the incomparable naturalist. I know that a number of generations back we are all related somehow – it is just a matter of statistics – but none of my friends or colleagues (apart from Sarah) has any direct link with Charles Darwin. It is difficult not to see this connection as a kind of blessing for the project – in the most secular meaning of that word, of course.

As we ambled through Lambridge Wood James and Monica explained that their father had been very much the conservationist until his death in 2003. Some parts of the wood (not ours) had been clear felled in rectangular plots, and replanted with conifers, mostly larch and Corsican pine. These are not natural trees to find in the Chiltern Hills; on the aerial view they show up as intensely dark-green areas quite distinct from the undulating beech crowns. The intention was to harvest the mature larch for pit props, but that project was evidently ill-conceived, since the Barlow ownership of the wood coincided almost exactly with the terminal decline of coalmining in Great Britain. Now, some of our fellow small wood owners are simply removing the larch to allow the broadleaved forest to recover. There is a great pile of conifer offcuts near the entrance to our wood; I decided to leave these fragments of mistaken forestry in order to study the processes of decay.

Elsewhere, the beech wood seems to have been left quietly to get on with being a beech wood, helped by periodic thinning. The manager of the wood was John Mooney, and the Barlows told me that they knew him as 'Eeyore' because of his pessimistic prognostications for making any money out of the wood. His annual accounts always finished with a thumping loss. It is as well that Barlow senior was primarily interested in good ecological stewardship, for all his correspondence from Mr Mooney is steeped in wry gloom.[10] The wood was under threat from trespassers, he said, or horse riders who cut barbed-wire fences, and poachers who poached. Deer of all species curtailed almost all regeneration, and what little was left was damaged by squirrels. The whole business was hamstrung by interfering busybodies and/or charlatans from the Campaign for the Preservation of Rural England and official bodies like English Nature. In 2000 his annual summary finished magisterially: 'It has been getting progressively worse for the past 25 years [before] hitting this nadir.'

Then Harry Potter came to the rescue. From 2001 onwards J.K. Rowling's novels about the young wizard were adapted for the screen, and the movies were watched by countless children. Many of them wanted their very own broomstick so they could play quidditch and generally fly about the place. The heads of the broomsticks were made from bound bundles of twigs, and the right kinds of twiggy shoots could easily be cut from birch trees and regenerating stumps in Lambridge Wood. More than a century ago there was an artisan known as a 'broom squire' who plied his trade deep in the beech woods, so it was a traditional skill.[11] Now there was an unprecedented broom boom, a besom bonanza. James Barlow said that they couldn't supply enough to the toy trade. In the end Lambridge Wood as a whole made at least a little money, in a thoroughly ecologically respectable way.

The distribution of trees today in our patch of Grim's Dyke Wood is likely to have been much the same when Sir Thomas bought the whole woodland, except that the beeches have had another forty years or so to increase their girth and height. Mr Mooney recorded that the wood escaped comparatively lightly from the great storm of October 1987, which flattened whole woods elsewhere (it didn't cheer him up). Many of the beeches are between ten and twenty paces apart, close enough to provide total leaf cover in summer, although there are several small clearings, and a large one on the northern edge where felling must have been more recent. Although beech is dominant, other kinds of woodland trees are a delightful addition to the silviculture. Eighteen magnificent wild cherry trees shoot skyward on sheer trunks to the same height as the beeches. Three stately ash trees decked in yellowish bark have spawned uncountable numbers of offspring. Less noticeable are wych elms discreetly hiding among the beeches. We have a total of just two oak trees, one of them a fine specimen, the other something of a poor relation, both tall. The same number of yew trees are the only conifers in our wood; these two are just at the beginning of their long, long lives. I scratched around for hours among brambles before finding a solitary field maple, and a tiny youngster at that, but I am glad to have it in my inventory.

Then there is the understory: trees of lesser stature that will grow happily in the shade of their towering neighbours. The most obvious is plentiful dark-green holly – probably too much holly. Still, I welcome it where its prickly evergreen foliage makes an almost impenetrable screen twice as high as a man around my favourite part of the wood: the Dingley Dell. Not quite in the middle of our patch, the Dell surrounds two of our most impressive old beech trees, which have been christened the King and the Queen. Unlike many of our beeches they don't soar away upwards immediately; there is a little spread of branches. Beneath these giants the ground is clear except for a covering of old leaves. Sitting on a log there in the April sunshine I feel as content as a dog before a fire. It is a place to write up my notes, and eat bacon sandwiches. Around the Dingley Dell a few old coppices of hazel – a traditional Chiltern undercrop – produce clusters of long, unbranched trunks almost straight from near the ground; these are of several ages and hence variable thickness. Some of the branches are dead – they need attention. A couple of young birch trees are growing on the edge of the large clearing. All these tree species have become old friends, and like all my friends they have quirks and history and several failings. We shall get to know them all.

Cherry blossom

During April the wild cherry blooms at the same time as the bluebells, but the cherry flowers are displaying high in the canopy. In hand I examine a flower head that has fallen down from above: coppery young leaves, half a dozen at the tip of the shoot all pointy and enthusiastic as if they should cry, 'Forward, forward!' But then behind this tip is a natural flower arrangement – ten little bundles of white cherry blossom coming off a grey-brown stick. They are arranged in clusters of four or five blooms, each one held on a green 'matchstick' an inch long. Every flower carries five notched, almost perfectly white petals surrounding yellow stamens, which are tiny threads with spherical heads like miniature pins (and in the centre of the flower, hardly grander, the style and stigma). Five red-brown sepals bend backwards from the flower as if to feign deference to the performance going on in front, which might be described as a cluster of tutus; and each bunch of flowers emerges from another five-fold arrangement of bracts next to the stem. So the twig is a series of bouquets topped by a flourish of leaves, a brief, exuberant festival of white blossom fifty feet above the common view. An early feast for insects, I suppose.

Why do we need those double garden varieties of flowering cherry – '*flore pleno*' and the rest? Admittedly they do augment the resemblance of the flowers to tutus, but there is already enough in the solitary blossoms. A Japanese artist might lose himself in a flower or ten: so short-lived, so fragile, like rice paper crimped into snowflakes. Even now a gentle snowfall of petals tumbling from high above is settling on last year's old beech leaves; in an hour or two the sun will have frizzled them into obscurity.

Butterflies appear suddenly in some numbers, and not just the umber-brown speckled wood butterflies, flitting erratically like camouflaged and subtle ghosts in and out of the shade, but also brimstone butterflies as bright and freshly coloured as primrose flowers. These last arrivals almost make up for the absence of real primroses in the wood, for the iconic springtime flower does not deign to live in the sparse, poor soil of Lambridge (this saddens me, as Darwin worked on primroses). A solitary peacock butterfly, a battered survivor of the winter frosts, with eyed wings shredded at their margins, is sunning itself on a bramble in the clearing, the better to gather the energy for a final burst of egg-laying. A green-veined white lingers for a second, then flits past and away.

I have evidently become attuned to the Class Insecta. In the midst of the bluebell sea, open flowers are pollinated by large bumblebees that delicately hang off pendent blossoms that look too frail to carry them. I fancy they are like oversize clappers hanging off the bells. I believe I can recognise the white-tailed (*Bombus lucorum*) and red-tailed (*Bombus lapidarius*) species, not least because they have a convenient dab of the appropriate colour at the end of their fuzzy abdomens. A huge red-tailed bumblebee must be a queen on the search for an old mousehole in which to establish a new colony. She buzzes about the cherry roots, and she won't have long to wait to find a suitable site. While I am crouching among the bulbs, a 'pretend' bumblebee whizzes past me that I know to be the bee fly (*Bombilius major*), one of nature's cruel deceivers. Although fuzzy and generally bee-like, it is no bee at all (it is closer to a bluebottle). It carries a long proboscis at its head end, and I watch it dart forward into and back out of a flower to feed on nectar, so it really is an entomological humming bird as much as anything. But it reproduces by laying its eggs near a true bees' colony, and its larvae crawl into the 'nest' where they consume the bees' grubs. In fact, it is an entomological Iago. I recall that Darwin described how deception was commonplace in nature; the man himself apparently so free of duplicity.

This is the day when all the male birds sing out passionately for a mate. Their plumage is buffed and preened, spring-ready. I am an amateur at birdsong, but I cannot mistake the sweet and penetrating phrase of the song thrush, repeated thrice or so, as if to emphasise its originality, for the next phrase is always different, and always repeated in its turn. I can pick out the implausibly loud song of the tiny jenny wren with a little rattle at the end of its performance. The songs of the robin redbreast and the blackbird I know well from my own garden. But I would not have recognised the nuthatch's broadcast had I not seen the handsome blue-backed bird sing from a bare twig: a kind of 'pwee-pwee-pwee' – simple and penetrating. Can it be that there is an inverse relationship between the showiness of the plumage and the beauty of the song? The nightingale and the most musical of the warblers are pretty ordinary of feather, while the extravagant peacock's raucous cry appeals only to other peacocks and the English aristocracy. Somewhere in the middle of this aesthetic spectrum, black-yellow-green great tits are everywhere in the wood uttering their repeated high regular notes – 'tee-too', possibly – which is hardly spectacular. The more sibilant, guttural, chatty conversation of the blue tit is more appropriate for such a small and bouncy cheeky chappie. Just now many blue tits hop rapidly about the denser branches whistling to one another, 'Here we are!' What I cannot do (pace the nuthatch) is reliably locate the source of all the birdsong; it seems to emanate in a general and celebratory way from almost everywhere. I begin to understand those descriptions of whole woods 'bursting into song'. The distant drumming of a woodpecker, a sporadic, hollow-sounding percussion, provides all that is needed for a back-beat to the avian orchestra. But then I briefly catch a glimpse of a timid tree creeper dodging behind a beech trunk, almost furtively working its way rapidly up the tree in search of insects tucked into tiny crannies in the bark that it can pick out with its curved bill. It moves in silence.

One cry is at odds with the general vernal celebration – a kind of wheezy, cross-sounding phrase repeated irregularly. Our pair of woodland buzzards are wheeling and gliding slowly round and round high overhead, as if barred from the general celebration below. Theirs is a simple call, almost like that of a baby working up to something more exciting. I had seen them yesterday flying through the wood itself: weighty, serious birds that appeared too substantial to negotiate their flightpath between the trees, something they nevertheless did with aplomb. Lambridge Wood is their patch. Beware, small rodents and unwary birds! If it turns out to be a good year for them, it will be a good year for the buzzards too.

Among the bluebells my eye is taken by something much more turquoise: a thrush's egg lying on the ground. It looks so perfect at first glance; a Mediterranean-summer blue overlaid with just a few black dots. I then remark a ragged hole in one side – somebody has taken it from its clay-lined nest and consumed the contents. The buzzards are exonerated (too much of the egg survives); I suspect a grey squirrel. I cradle the empty shell in the middle of my palm. It is almost impossibly light. Surely this must be the first item for my wood collection; I must cherish it.

And then my eye is caught by a perfectly white bluebell, just one among so many thousands of the common kind. I suppose it should be called a whitebell. It is as rare as a sober Irishman on St Patrick's Day, but much more conspicuous. It stands out from the crowd, visible yards away. It is the result of a natural mutation. If it were a successful mutation I suppose there would be many more of them, but there it is, living proof of that molecular part of the science of evolution that Charles Darwin did not know about. Just one tiny change on the DNA code and blue becomes white. Since most bluebell reproduction is from the proliferation of the bulbs, if I had a mind I could lift this example, nurture it in my garden, and artificially ensure its success. I could call it variety 'Grim's Dyke'. The origin of so many white garden flowers is thus revealed: white campanulas, that are so blue in the hedgerows; white pinks (never, after all, 'pink whites'); white honesty; even white pelargoniums. Like the wild cherry, some are born white; others have whiteness thrust upon them.

Ground elder soup

The first ground elder shoots (*Aegopodium podagraria*) are prolific near the edge of the wood. This plant is a notorious garden weed that, once established, is almost impossible to eradicate from the herbaceous border, but in a wood it makes a prettier sight and a more constrained patch. Its lobed and divided leaves appear well before parsley-like flowers. When the leaves are young and pale green, I discovered that they are a good vegetable; they become rank a month later. So there is a different way to view ground elder: as food! Ground elder soup is simple to make. A bagful of young leaves is gathered easily enough. The coarser stalks must be broken off, and the leaves

Half-timbered cottage in the village of Assendon closest to the wood. This cottage was the home of the writer Cecil Roberts during the middle of the twentieth century.

Spring view of Lambridge Wood, by Rob Francis.

are roughly chopped. A finely-sliced onion is softened by frying in butter, until it just starts to caramelise. At this point a medium-sized floury potato is added, chopped into small cubes, and placed with the leaves and onion in a heavy pot, and then a generous quantity of stock (or 1½ to two pints of water and a chicken or vegetable stock cube) together with a pinch of mixed herbs and pepper to taste. After it has been brought to the boil it is simply a matter of simmering over a low heat until the potato is soft, when the whole can be blended in a liquidiser. Croutons or a swirl of cream add a finishing touch. I should say that there are other wild members of the parsley family that are poisonous, most particularly hemlock. There should be no risk of mistaking the feathery leaves of hemlock for the rose-like leaves of ground elder, but if in doubt leave well alone.

2

May

First felling

It has been raining for several days, but there is still not enough closed canopy aloft to provide any kind of shelter. The beech trunks are sodden, and now also distinctly green: rainfall has woken up tiny algae and liverworts living on the bark, and they are growing rapidly while they can. A large log near one of the paths has been rotting away for years, and what remains of its wood absorbs water like a sponge: maybe it was a standing cherry ten years ago. A bright-yellow lobe is growing out from one side of it, an excrescence both luminous and unnatural in its brilliance, like a glowing and irregular ox tongue. Every day it seems to add another inch or so, as if licking itself into further substance. The fungus is feeding on the wood: I know it as the fruiting body of the sulphur polypore (*Laetiporus sulfureus*).[1] I have seen it on several trees, but it does have a common preference for wild cherry (*Prunus avium*). When it is further developed I know it will become more like a bracket, and on its underside hundreds of tiny pores will develop, marking the ends of tubes in which its spores are produced in their millions to waft away on the lightest breeze, randomly seeking out the perfect tree on which to germinate and prosper anew.

This damp period favours natural succulence – living things that are full of juice. On another part of the log three or four bright-pink-coloured balls are the size of small children's marbles. They too look unnatural, like dropped beads of coral that have no place in a beech wood in England in spring. Prodded with a finger, they burst like boils, spattering pink juice. My daughter hates them, despite my protestations that they have a weird beauty. They are the reproductive spheres of a plasmodial slime mould, *Lycogala epidendrum*.[2] As its common name implies, it was once thought of as fungal, but it is not a mould of any kind, though the sliminess cannot be gainsaid. Today the balls are forming everywhere in groups on the woodpile near the barn, dozens of them. They thrive in the damp. For the earlier part of their life cycle they moved along and through the forest floor, like amoebae, in a subtle but bounded transparent body with thousands of nuclei, where they soaked up nutrients from decaying organic matter. If my daughter were to say they were creepy at this stage, that would be no less than the truth. They creep and they grow. When they have grown enough – and it is an interesting question just what it is that says, 'Enough!' – they change character more thoroughly than did Dr Jekyll to Mr Hyde. They glide up to a higher piece of dead wood to turn into those pink balls. At this stage the transformation is incomplete, but in a week or two the balls will have turned brown and become much less conspicuous. A few weeks later they will have transformed into masses of umber-brown microscopical spores – dust, to the eye – and will then be blown to the four winds. On another piece of wet wood I discover a weft of tiny, white, delicate gelatinous fingers hanging down like stalactites: it is another 'slimer' (*Ceratiomyxa fruticulosa*). There's a sudden vision of the wood as a mass of almost invisible cells sliding and questing through the dampness.

The sun returns at last, and with it a gentle breeze. The weak solar rays pick out the fluttering foliage of freshly unfurled beech leaves in the softest shades of pale green – almost yellow in a certain light. On the ground lie hundreds of tiny brown bracts that had encased the nascent leaves over winter in thin, spiky buds. Now they are redundant. I examine a new beech leaf under a lens: it is fringed with white hairs more delicate than a baby's eyelashes. It has not yet acquired anything of its summer rigidity; it is like tissue paper. On the low branches of the trees the leaves quiver gently, making tiers of light thrown into contrast against the unchanging dark of the holly. It is almost as if we were under water, and the leaves were being stirred by invisible currents. Where the sun sneaks through the forest to illuminate the cherry trees the polished surfaces of their bark shine almost silver.

Cousin John is felling a beech tree that is leaning dangerously over the public footpath. Accidents in woods caused by falling branches are very rare – most people have the sense to stay out of the woods during tornados. But beeches sometimes shed a whole branch just for the hell of it: these are called 'widow-makers' (they never fall on girls). We cannot have that happen to a passing dog-walker. John starts with a saw on a long pole to trim off the side branches. One of them almost touches the ground, such is the curve on the tree over the path. He can reach more branches with a chainsaw borne aloft on a staff; then, with a bigger chainsaw, the main action. The saw makes a raucous, grating racket, with something of the unforgiving persistence of the dentist's drill about it. John is a professional, so he sports massive earmuffs and mighty gauntlets. Bystanders are reduced to making encouraging gestures and gurning amiably.

There is only one way for the tree to fall, but there is a skill to making the cuts so that it does not spring any surprises. The poor beech groans, crackles like a fusillade of fireworks, and then it is down, just like that. With a girth of three feet there is a lot of firewood to be mined. Despite its lopsidedness the tree is still very much alive, so now on the ground its branches stick up all stiff and unnatural, decked in new leaves that flutter in the breeze for the last time. They will be limp by next morning. The centre of the trunk proves to be quite rotten – black, and hollowing out. The tree would have cracked eventually, so it was as well we took it down. The fungal damage extends further up the trunk until it is visible only as a curious kind of dark hieroglyph in the centre of the log. John cuts lengths from the upper part of the tree that will be taken away in his van to be sawn into rounds, then split into logs for next year's open fires at home. The fat end of the tree is winched into a position where we can use it as a rustic seat in future and study its slow decay. The brash is carried into the wood to rot away and return to the soil. Who would believe that one, not particularly large, tree could generate so much work?

Falling beech branches have crushed a few bluebells, but no matter, they are already showing signs of decline. The great sea of blooms has deepened to an azure colour, and still looks unbroken from afar, but the lower flowers on the spikes are already blousy and fading. Their dark-green blades of leaves have lost the vigour of their youth, and have started to become flaccid and unenthusiastic. But as one flower starts to fade, another prospers.

Patches of ground near the bluebells are covered with neatly tiered rosettes of lance-shaped leaves. A cluster of tiny white flowers crowns each tier; this plant is the perennial sweet woodruff (*Galium odoratum*), no taller than the bluebell. Each flower in the cluster has only four wee petals; under my lens the edges of the leaves can be seen to be lined with minute prickles that can be felt by gingerly stroking a finger along their margins – usually eight delicate leaves to a whorl. As for the sweetness implied in both the common and the Latin name, this little plant has a fragrance more persuasive than that of any bluebell. Its sweetness grows as the woodruff is dried – which it does readily. I placed a bunch in a warm airing cupboard and it was preserved in a day. Dried woodruff was once slipped between the sheets to sweeten bedding, so the sleeper might dream of woods in spring. The scent of new-mown hay is the same chemical agent (coumarin) as in woodruff, but only if that hay has sweet vernal grass as one of its components. Sweet woodruff is used to flavour traditional springtime drinks and sweets in Germany, though its German name of *Waldmeister* ('master of the woods') seems too assertive for such a refined little plant. My son thought to try it as a 'botanical' to flavour gin; there has been a revival in boutique spirits, and he wants to be ahead of the *Zeitgeist*. Although juniper berries are the traditional flavoring for gin, all kinds of refinements are possible by steeping other herbs and spices in the spirit. Experimental Batch Number One was rather overpowering, but Experimental Batch Number Two turned out to be delicious. It was presented with a tasteful label featuring the wood in spring. Grim's Dyke Gin may yet feature in some future genre market.

Bluebell and sweet woodruff are specialists. Not every bulb or herb can thrive in our beech woodland. Timing is all. These plants have to steal as much light as they can before the canopy shuts off the sunshine. There is really no option but to flower in spring. They join the dog violet and the lesser celandine in the early shift. Wood melick, which makes something approaching a field of lively green over parts of the wood in May, will produce its nodding rod of single flowers and then fade before summer is over. The bluebells' burst of photo-synthetic activity is done even before the beech leaves mature, and the energy the plants have gained during their brief but glorious exuber-ance is stored in the bulb. Job done, everything above ground withers away. The lesser celandine's[3] pretty, heart-shaped leaves also enjoy but a brief existence; they soon turn yellow and shrivel. They too seques-ter energy in little cream-coloured, bulb-like storehouses that linger in a somewhat scrotal cluster below the ground through most of the year. I have seen similar-looking bulblets form at leaf junctions, any one of which might produce a new plant next season. The leaves of sweet woodruff and violets linger on, a little dowdily, after flowering, but the woodruff has a strong root network that can survive a major drought unharmed. I add two more pleasing spring flowers discov-ered from snuffling around in the wood: a discreet purple-flowered wood speedwell, *Veronica montana*, modestly creeping along a path-side where there is a little more light, and a pretty buttercup, goldi-locks (*Ranunculus auricomus*), with apparently much the same requirements. Both have the kind of understated beauty that rewards a little botanical nous.

Our slowly spreading English bluebell is a marker for ancient woodlands. Stately wood spurge (*Euphorbia amygdaloides*) tells the same story, and I have found it in four places. It almost makes a small shrub. Its young, leafy shoots are a lovely coppery hue and slightly pendent, contrasting well with the flowering heads, which are a lively green. All spurges have peculiar flowers, unlike those of any other plant. They have none of the usual paraphernalia: no petals or sepals, and the reproductive parts are reduced to the minimum. What could be mistaken for petals are actually yellow-green leafy bracts that form a kind of cup around the minimalist sexual business. A cluster of these distinctive structures makes the flower head. Unlike lesser celandines and bluebells, the spurge plant stays on in the wood, gradually losing its fine vernal contrasts, and fading with dignity. I have seen spurge species thriving in deserts looking just like cacti, and others creeping on seashores, and yet more growing far too vigorously in my vegetable patch, so it is scarcely a surprise to find one species that likes to live in old Chiltern beech woods. All spurges carry a horrible, poisonous white sap that seeps out if a leaf or stem is broken. Once I accidentally rubbed a minute amount of the milk into my eye and danced around for two hours in excruciating agony, weeping profusely. I cannot recall such a painful reaction since they closed my favourite Chiltern pub (the Dog and Badger).

Men of letters

Writers are not a rare species. They seem to crop up everywhere, rather like spurges, although some are less poisonous. I confess that at first I thought I would have my patch of Chiltern Hills beech woods to myself. I was wrong. Over the brow of the ridge behind Lambridge Wood Barn, in the village of Lower Assendon, and just beyond the Fair Mile leading out of Henley-on-Thames in the Oxford direction, a small Tudor cottage decked in oak beams was home for several decades to a famous writer: Cecil Roberts. In the 1930s Roberts published a series of three books centred on Pilgrim Cottage: *Gone Rustic*, *Gone Rambling* and *Gone Afield*. I now have them all in hardback, though had I not bought Grim's Dyke Wood I would probably never have heard of this particular author. *Gone Rustic* was reprinted at least six times: it was a bestseller. All the books are charming, gossipy, name-dropping confections about life in a kind of idealised Rustic. Beneath the dustjackets they have *bas-relief* covers with cottagey timber framing built in. Roberts's is the same world as that in which Hercule Poirot joined genial house parties in small stately homes only to find His Lordship dead in the drawing room. It has an exact fictional match in E.F. Benson's Mapp and Lucia novels, set in a genteel part of Sussex where private incomes would pay for house-keepers and cooks, and the protagonists could concentrate on paint-ing watercolours and choosing chrysanthemum varieties. Working-class country folk tended to have only colourful walk-on parts.

According to Cecil Roberts's account in *Gone Rustic*, he discovered Pilgrim Cottage in 1930 by accident after sustaining a puncture on the road from Henley to Oxford. He writes: 'Around me the view was imposing, almost Tyrolean, with steep larch covered hillsides, and in the distance between thick beech woods nobly clothing the green-sward, a ravine.' The last may have been a reference to the gentle valley leading to Stonor. Pilgrim Cottage is still much as it was in the middle of the twentieth century. Cecil's upstairs windows would have commanded a view of Lambridge Wood on the near skyline, so he really *was* a neighbour. I imagine him fussing around his garden, absorbed in his gladioli, while instructing his housekeeper to lay tea for the Marchesa, who would be arriving betimes in the Hispano-Suiza. Pilgrim Cottage, he complained, was positively *deluged* with visitors, all of them fascinating, making the necessary wielding of his pen a matter of some concern. In spite of all his socialising, he did manage to produce a quantity of books and much verse. The core of his Pilgrim Cottage books is provided by local history, well described, and much of it relevant to my story; and his tales of local craftsmen are invaluable. His other love is Italy, and he jumps to Venice and palazzos and the story of the Finzi-Continis at every hint of a meta-phor. His interest in natural history was as perfunctory as his interest in gardening and sunbathing was obsessive. The Chilterns provided a green backdrop to his real concerns, which were always human.

Cecil Roberts had another life during the Second World War. Pilgrim Cottage and its stories had a great following in the United States. Their appeal may have been rather akin to the current popu-larity of sagas featuring big houses and their goings-on a century ago. Roberts was recruited to aid the war effort by giving lecture tours in

America, which he evidently did with great success. The *New York Times* reported that 'the best propaganda in the world is the British and the most efficient expression we witnessed were the lectures held all over the USA by the noted author, Cecil Roberts. These lectures never had the flavor of propaganda but brought more good will towards Britain than anything else.'[4] His charm could obviously do its work far away from the Thames Valley.

When the conflict was over Roberts felt he had not received sufficient official recognition for his efforts. He tried to restore the balance by publishing his autobiography in no fewer than five volumes. Rowena Emmett, daughter of Mr and Mrs Plater, the next occupants of Pilgrim Cottage in 1953, told me that for many years beaming Americans would turn up at their garden gate requesting permission for photographs. 'It was,' she said, 'quite a nuisance.' The former owner did not downplay the fame of Pilgrim Cottage. He had written to the Platers advising, 'many thousands all over the world love it, for it symbolizes England for them'. When Rowena met Cecil Roberts she was a young schoolgirl, and she found him more than a little alarming. She later realised that his manner was just very camp. A modern reading of *Gone Rambling* would leave little doubt about the author's sexual orientation, with its panegyrics to bare-chested Italian sawyers whose 'skin, tanned a warm mahogany by the Venetian sun, gleamed and caught a hundred tones and facets of light as the muscles glided with cryptic strength beneath their satin sheaths'; to say nothing of some very ambiguous poems.

I cannot help wondering how Cecil Roberts might have been regarded by the locals in the Golden Ball in those less tolerant times. Maybe they just thought he was from London. He seems to have been genuinely helpful with his time and money in the village, and nearly always described its inhabitants sympathetically. He appreciated the efforts of his gardeners, including Charles Crewe, who lived in a damp dwelling with none of the usual services on the very edge of Lambridge Wood. Roberts was determinedly anti-fascist. I was surprised to learn that his origins in Nottingham were far from aristocratic. He was a self-made man, living mostly from his prolific pen, whose name-dropping was probably an exuberant validation of his reinvention. *'Mon dieu!'* his special and amusing spinster friend Miss Whissitt might have exclaimed. *'Tu es une arriviste!'*

H.J. Massingham is altogether more astringent. His 1940 book *Chiltern Country* deals with the whole range of hills in luminous language. His feel for natural observation is superb. Much of his work is driven by fury about the spread – no, the rash – of homely villas outwards from London. He mourns the 'real England, the England in which the hills, the vales, the waters, the crops, the roads, the buildings, the natives and the rock that bore them up all on its back were intricately bound together in an organic system not unlike the human body'.[5] The country cottages that have withstood the centuries – and the worthy souls who have earned their living around them for as long – are becoming overwhelmed by red-brick mediocrities planted about with shrubs that don't belong. Beech woods become desirable scenic accessories rather than essential resources. For Massingham the country beyond Rickmansworth was irretrievable, and the country around High Wycombe was doomed. The spread of the Metropolitan Line from London into the hills was a sinister fungus that sprouted despicable edifices – suburbia: 'the touch of it annihilates identity in place'. His ruralism stands at the other extreme to the poet John Betjeman's sympathetic regard for what he termed 'Metroland', a land of healthy young women and clean semi-detached gentility.

Massingham's combativeness is quite appealing. I think he would fain have jumped back in time way past the Enlightenment to fetch up somewhere in the late medieval period. He reserves his most eloquent writing for our piece of country, and most particularly Stonor Park, 'the heart of the Chilterns', where the wild spirit of the place has not yet been ousted, the views not hopelessly corrupted with eyesores. I have no proof that he ever visited our woods, but I hope he would have found the *genius loci* satisfactory there, too. I am certain he would have disapproved of the practice of 'splitting' to sell off ancient woodlands, thereby dividing the integrity of manors that had been in existence for nearly a thousand years. There is no defence, except to say that I could never have afforded to buy a whole stretch. There are plenty of very wealthy people in the hills who don't appreciate the unique treasures they have on their land, and my small patch is much loved.

Just over a century before Cecil Roberts was pottering around his garden in Lower Assendon, John Stuart Mill was exploring our Chiltern countryside with a far more scientific enthusiasm. The philosopher and political theorist was equally a dedicated and scholarly botanist. Very few people can instantly recognise rare plants like wintergreens (*Pyrola*), but J.S. Mill was one of them. From his early days he was a close friend of George Bentham (nephew of Jeremy), who would become one of the greatest botanists of the Victorian age. Mill made an expedition in France in search of poorly known flora. His house in Kensington Square in London was virtually a herbarium. Some people who are not naturalists find it odd that famous thinkers, poets or mathematicians might derive as much pleasure from the minutiae of natural history as from the fields of endeavour that made them famous. Vladimir Nabokov was as serious about blue butterflies as he was about writing novels, but certain critics relay this fact as a kind of eccentric footnote to the life of the artist. Doubtless they perceive that less time frittered away with the butterflies might have resulted in one or two more novels. Can they not see that the taxonomic eye applied to recognising the subtlest nuances of difference in butterflies is the same eye that spots the deceptions and evasions in human motivation? The capacity to make accurate observations is a special genius, and it is not limited to focusing on one particular bipedal subject species.

In 1828 J.S. Mill undertook his own bipedal tour that passed through our part of Oxfordshire.[6] Open fields yielded abundant white-flowered wild candytuft, 'one of the commonest of all weeds' (*Iberis amara*), which is now a rare plant – I eventually ran it down myself on clear ground on Swyncombe Down, nine miles from our wood. His record of thorow wax (*Bupleurum rotundifolium*) might be one of the last for the county: this species is close to extinction in Great Britain, and Mill noted its rarity even then. On 5 July he approached Henley from Nettlebed: our patch. You may imagine the pleasure his subsequent writing gave me.

> The woods are the great beauty of this country. They are real woods, not copse, that is, they are not cut down for fire-wood, but allowed to grow into timber, though not to any great age, nor are there, as far as we could perceive, many very large or fine trees among them … We stopped at the White Hart, Nettlebed for the night, and in the evening walked down the hill by the Oxford Road towards Henley. It passes through a fine forest-like beech wood, and on the whole the ascent to Nettlebed from Henley is far more beautiful than any thing else which we have seen in its vicinity.

I cannot prove that John Stuart Mill walked *exactly* along the footpath past our wood, although it is hard to see how a woodland ascent towards Nettlebed from Henley could have taken any other route. His praise for its beauty is not the least of it. The woods he described are very like those that still flourish today in this corner of the Chiltern Hills; the same stately 'forest' of mature timber trees, but yet lacking any truly ancient giants such as survive in old parkland or as parish boundary markers. My wife and I discovered a massive ancient beech pollard along a path in Nettlebed that must have been four hundred years old at least, all gnarled and knobbly and hollowed out. There are a few in the area. But no, Lambridge Wood was a working wood two hundred years ago, a beechen grove permitted to grow on to timber, but not to senility. We shall see, however, that nothing is forever, and our wood would have had different employment in earlier and later times.

Nor can I prove that John Stuart Mill walked through the wood in the company of George Grote, but I like to think that circumstances favoured it. They were friends already in the early 1820s. Mill was both an admirer and a reviewer of Grote's writing, and particularly his monumental history of Greece (1846–56) in twelve volumes:[7] a work not perhaps as beloved as Gibbon on Rome, but with a similarly vast reach and ambition. The two prolific writers shared what might broadly be called liberal and reformist views, and were Utilitarians. The seat of the banking Grote family was Badgemore House, which has been mentioned as the estate adjoining Greys Court directly to the east. Part of the Henley end of Lambridge Wood was within that estate; tracks ran onwards into our part of the greater wood. George's father was fond of country pursuits, and went hacking on horseback through our woods and onwards to Bix. Young George (then still a banker) and his wife would make the forty-mile journey from London to spend ten days with his parents, and on one occasion Mrs Grote drove all the way in her own one-horse vehicle while her husband rode for four hours separately on horseback.[8] Although George Grote was much attached to Badgemore, in the days before the railway it was hardly a practical commute. By 1831 it was clear that the country house should be given up, and George left for the metropolis to devote more time to reformist politics. We shall see that all the manor houses surrounding our wood had political connections with the capital at one time or another.

As for contemporary writers, Richard Mabey's memoir of Chiltern countryside[9] is centred on a region not very far from Lambridge Wood, while Ian McEwan described a long walk through our chalk country in his 2007 novel *On Chesil Beach*. This area of southern England proves to be almost as crawling with writers as with other invertebrates.

Hard grounds

The ground in this part of the wood is crunchy under my boots. Beneath a few of last year's fallen leaves and under the questing loops of bramble shoots there appears to be nothing but rock. I am attempting to dig a hole to explore the surface geology, but my spade refuses to make any progress. Its blade twists and complains against a barrier of stones. I will have to employ my geological hammer to solve the problem.

The pick side of the hammer starts levering up lumpy flints, some bigger than my fist. They leave the damp ground reluctantly, with a sucking noise. Where I hammer downwards into the growing hole, sparks fly where steel meets flint. Briefly, there is a smell of cordite; in the days of flintlock pistols that smell would have been a familiar one. Flints were used to strike the spark that ignited gunpowder before a shot could be made. Our flints are embedded in reddish ochre clay that tries to hold on to them, clay that can easily be rolled into a coherent ball between the palms of my hands, and sticks to the fingers. The exterior of most of the flints is white when wiped clear of its clay coat, but where the hammer has shattered one of the larger flints its interior is strikingly black, and mottled in patches. It is a hard rock, but a brittle one shot through with flaws. Much of the wood is effectively floored with flint. Of the chalk of the Chilterns there is no sign.

Just down the hill beyond the Fair Mile I know that chalk underlies everything. When the dual-carriageway road was repaired great masses of the white rock were dumped on the side, and I picked out a typical, conical fossil sponge called *Ventriculites* from the rock pile. Even within Lambridge Wood, further downslope towards Henley, a mysterious excavation known as the Fairies' Hole (marked on even the oldest maps) is undoubtedly dug within the white limestone. The rock that makes the whole range of hills, 'the rock that bore them up all on its back', as H.J. Massingham said, is an understory of chalk. Within the chalk, hard flints form discrete layers, but they never dominate completely. This flint was ultimately derived from fossil sponges within the chalk that had internal skeletons made of silica struts. The silica was first dissolved, and then re-deposited in flinty layers as the original chalk ooze gradually hardened and transformed into the rock we see today. Whatever underlies Grim's Dyke Wood on the higher ground evidently also lies on top of the chalk formation, but is largely made of flints derived from it, all stuck in a matrix of sticky clay. This deposit is called, unsurprisingly, clay-with-flints, and in the wood the flints are dominant.

Clay-with-flints caps the chalk in many parts of the Chiltern Hills.[10] It is the product of many millennia of slow solution and weathering-away of the chalk; it is what is left behind when everything else is removed. Chalk is weakly soluble in rainwater, which is why water derived from an aquifer in the Chilterns leaves a limescale deposit behind in a kettle. After a very long time, as the chalk naturally disappears the originally scattered flints become concentrated. Flint is insoluble; in fact, this form of silica is well-nigh indestructible. It can be tossed into rivers or buried in gardens for centuries, and emerges unscathed. It will outlast the Chilterns.

To estimate the thickness of the clay-with-flint capping I walk slowly up from the end of the Fair Mile to Lambridge Wood along Pickpurse Lane (see comments on highwaymen, pp.209–12), digging with my hammer into the bank until the telltale milkiness goes out of the soil. There are other signs to look for. Old man's beard (*Clematis vitalba*), the nearest thing in the British flora to a liana, only grows on chalk – it will not tolerate clay-with-flints. Wild marjoram is no more forgiving. Plant roots sense chemistry with the exquisite palate of a connoisseur. Both the indicator plants grow in abundance near the bottom of the lane and fade away upward. By the time all evidence of chalk has disappeared I conclude that very roughly twenty feet of clay-with-flints must lie above. That is sufficient to make the thin soil on the high ground neutral or acidic compared with the alkaline soils on the slope and in the valley bottom. This saddens me, for many of the more glamorous plants love chalk: the whitebeam tree with big simple leaves with shining undersides; cheerful yellow St John's wort; and many an orchid. I shall just have to live without them – I cannot argue with geology. Now I also know why our footpaths can become like quagmires after too much rain. That layer of impermeable clay does not drain well; it likes to make ponds. Some corner of our wood will always be damp.

Back to Grim's Dyke Wood. I decide not to try to excavate much more of the recalcitrant stony ground. Instead I shall use the holes I have made to put down beetle traps, burying a few cups half-filled with lethal Dettol to ensnare night crawlers. As I tidy up, a different stone surprises me. Lying on top of the ground by a beech trunk is a pebble the size and shape of a goose egg. It is purple, and it is certainly no flint. Under my hand lens I recognise it immediately as hard sandstone. I soon see more examples of a similar cast, liver-coloured, always rounded off to make satisfactory hand specimens, by which I mean something that sits easily in the palm. They are all strangers. There is no rock formation I can think of in the Chiltern Hills, or in the Vale of Aylesbury beyond, or even further afield beyond Oxford, that might produce such pebbles. They have all their corners chipped off until they are satisfyingly elliptical in outline, and smoothly rounded at the corners. This is a form sculpted by long sojourn in a lively river; erosion has knocked them into shape little by little, polishing repeatedly over a very long time. How could they have got here, into the middle of our beech wood? There are other strangers too. A white pebble that might be a pigeon's egg, judging from its shape and size; it's another form of silica – resembling flint, but with a dense, swirling milky whiteness. Vein quartz, I will wager. It might have originated from a vein within granite or snaking along a fault fracturing other rocks. There is no source for such vein quartz anywhere around here. Strewn on top of the clay-with-flints are a bunch of lithological vagabonds from afar.

I decide to investigate further. At the Natural History Museum a skilful colleague cuts sections through my errant pebbles. Microscopic examination should show what they are made of, and reveal the secrets of their derivation. The samples are sliced using a diamond saw; then a thin sliver is mounted on a glass slide and reduced in thickness so much that light can penetrate the minerals that make up the rock; they can now be examined under a petrological microscope. I learned my microscopy skills as an undergraduate in a dusty laboratory in Cambridge, and distant memories stir as I stare down the eyepiece.

The vein quartz pebble proves to be typical. Under the microscope it shows as an irregular patchwork of grey or slightly yellowish crystals, with trails of tiny bubbles. It could have originated from several geological sites. However, one sample has several good pieces of similar-looking rounded vein quartz embedded *within* a chunk of the sandstone, like plums in a pudding. Maybe the quartz pebbles were derived from the same sandstone formation, only a part of it that was much coarser – a conglomerate, in geological terms. The pebbles must have been incorporated into the sandstone from some still older source. The sandstone itself is curious and distinctive. The individual sand grains are clear enough as masses of rounded outlines under the microscope, and they are of similar size to those that might be found on a beach today. But they are glued together by dark-red cement, without doubt full of iron. This is the mineral that gives the pebbles their rich red colour. The sandstone is recognisable, and it can be run down to its source. The pebbles must be Triassic in age (about 235 million years old), and they come from the English Midlands.[11] The old name for them was from the German – *Bunter* sandstone[12] – and they date back to a time when Britain was hot and arid and the geography of Europe had an utterly different cast. As for the indestructible milky quartz pebbles, some of them originated from the erosion of still older rocks long before they in their turn became incorporated into the *Bunter* sandstone; they might be as old as a billion years. Enmeshed under our own beech roots we have pebbles that account for a quarter of the history of the earth; and they arrived in the Chilterns by water, without question.

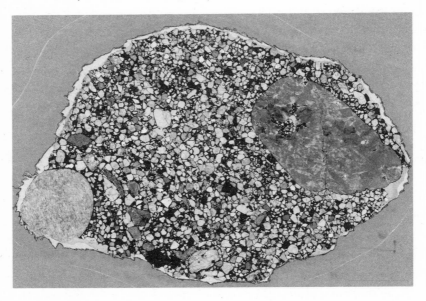

Thin section of a conglomerate pebble: an alien in the wood.

The vigorous river that brought down the pebbles from eighty miles to the north-west was an ancestor of the same River Thames that now flows sedately two miles to the east of the wood.[13] During the Pleistocene Ice Age (2,588,000 to 11,700 years ago) thick continental glaciers to the north waxed and waned by turn, diverting all Europe's great rivers at some times, providing the source for vast spreads of gravel at others. The ancient Thames left behind a record of this complex history in its former river terraces, the remains of which are scattered around the Chilterns and the London Basin. The oldest of these terraces is close to our wood, at Nettlebed. The exotic pebbles that I found in the wood are well known from a younger terrace, a set of strata called the Stoke Row Gravels.

The village that gives that formation its name is about four miles west of the wood, high on the Chiltern plateau. It is home to a most implausible structure, a little piece of India by a village green such as Cecil Roberts would have described as being quintessentially English. The Maharajah's Well was dug by hand 368 feet down into the chalk, passing on the way down through the overlying gravels relevant to our wood, and all at the personal expense of the Maharajah of Benares, who also supplied the exotic, elegant and ornate canopy. His gift was reciprocation for a well dug in India at Azimghur by Edward Reade ('squire' of Stoke Row) in 1831. The Maharajah remembered that Reade had told him how his little home village on the top of the Chiltern Hills was most precariously supplied with water. His remark-able gift of the Maharajah's Well was officially opened in 1864, and did its job efficiently for seven decades.

Professor Phil Gibbard tells me that the Midland 'connection' was open for well over a million years, until about 450,000 years ago. Although the huge Pleistocene ice sheets never reached as far south as the wood, their influence could not have been more profound. An icy climate sculpted the Chiltern landscape. It scrubbed the landscape to a *tabula rasa* on which all its subsequent history was inscribed; this marks the baseline of my natural history. I have to imagine a land-scape stripped of trees. The slopes of the hills are bare, with only the hardiest herbs able to cope with the frigidity to the south of the permanent ice. Now indeed Cecil Roberts's description of the valley up to Stonor as a 'ravine' may be nearer the mark, for the Chiltern country is riven with steep-sided valleys. Cold summer streams that flow with rejuvenated force following the annual melt carve vigorously down into the soft chalk, which is still too deeply frozen to allow the tumbling waters simply to be absorbed. The streambed is choked up with flint pebbles. In Arctic latitudes I have watched just the same fitful progress of jostling stones during the brief summer – their percussion kept me awake. The legacy of the frozen era still marks the ground: not only the implausible sheerness of some Chiltern hillsides, but also valley bottoms floored even now by ancient stream gravels.

Old names were bestowed by the Ice Age, like Rocky Lane, which runs up a valley on the south-western side of the Greys estate. Then, somewhat over eleven thousand years ago, the climate warmed for good, and now I must populate the hills with trees. Pioneers at first, small willows, hardy conifers; then birch, pine and aspen; and next, and not necessarily in this order, the broadleaved trees that came to make the original wildwood: oak, ash, lime, elm, hazel and beech. Oliver Rackham[14] tells us that the lime species he calls pry (*Tilia cordata*) – the small-leaved lime – was dominant in many of those early woodlands. It still lurks, mostly unremarked, in a few places in the Chiltern Hills, but not in our wood. About six thousand years ago 'Stone Age' humans were already beginning to fell the virginal forests, where previously arboreal old age and accident had been the only foresters. The streams that had once carved the 'ravines' were now absorbed into the defrosted chalk, leaving a legacy of steep dry valleys, like the one that runs from the Fair Mile to Stonor Park; though it is not *quite* dry, for after unusually wet winters the water table rises until streams such as the Assendon Brook reappear, bounding alongside the tiny roads and causing cyclists to swerve and walkers to chide their wet Labradors.

I hold a couple of the liver-coloured sandstone pebbles and a quartz keepsake up to the May sunshine. So much can be read from these fragments. I think of the lines from *As You Like It*:

And this our life, exempt from public haunt,
Finds tongues in trees, books in the running brooks,
Sermons in stones, and good in everything.

These remarkable, sermonising samples of rocks that might have passed unnoticed are next to be added to the collection.

Maiden ladies and geraniums

In 1787 Mary, Dowager Lady Stapleton, moved into Greys Court as her dower house, and women dominated that establishment for the next eighty years. After she died at the age of ninety-one in 1835, Mary's daughters Maria and Catherine stayed on in the big house that owned Lambridge Wood until the younger sister Catherine's death twenty-eight years later; both sisters also lived to a great age. The intellectual ferment in London that preoccupied their neighbour, George Grote – and the circle that included John Stuart Mill – passed them by. Rather, the Church engaged them fully, and led them to charities directed at the moral and religious education of the less fortunate in the parish of Rotherfield Greys. The rents from tenancies guaranteed their gentility, if not their spinsterhood. It must have been a quiet time at the ancient house.

Mary's son James was at Greys Court in the earlier days, and his friend from Christ Church, Oxford, Charles Kirkpatrick Sharpe, stayed with him often, and wrote frank letters to his mother at Hoddam Castle peppered with observations that exactly match his surname.[15] On 12 January 1801 he was describing his Christmas at Greys, 'which began, woe's me! like most other gambols, with laughter, and ended in tears'. He described the entertainments the local town had to offer thus:

Miss Stapleton, her brother, and myself, repaired in high feather to a ball at Henley, the night after Christmas, and were much amused in many ways. The company consisted of the town gentry, and the progeny of farmers in the neighbourhood; the clowns with lank, rat-tail hair, and white gloves drawn tight on hands which they knew not how to dispose of; the clownesses with long stiff feathers stuck round their heads like those of a shuttle cock, and wealth of paste beads and pinch-beck chains. They came all stealing into the room as if they were doing some villainy, and joyful was the meeting of the benches and their bums. But the dancing did them most ease; the nymphs imitating the kicking of their cows, the swains the prancing of their cart horses. But joy of joys! Tea was brought at twelve, and off came all the silken mittens and pure white gloves in an instant, exposing lovely raw beef arms and mutton fists more inured to twirl mopsticks and grasp pitch-forks than to flutter fans or flourish bamboos.

There is a precision of observation here that almost mitigates the snobbery. Walter Scott wrote of Sharpe: 'he has great wit, and a great turn for antiquarian lore'. Nor did the poor Misses Stapleton escape his gimlet eye. A year later he wrote:

> I made out my visit to [James] Stapleton, and yawned with him for a week. They are such good dull people at Greys Court! The sober primitive women do nothing the whole day but fiddle-faddle with their greenhouse, like so many Eves, and truly they are in little danger of a tempter, for their faces would frighten the devil, not to mention men.

The only portrait I know depicting the sisters (and brother), by Thomas Beach in 1789, suggests this judgement might be unfair. The large painting hangs on the staircase in the grand Holburne Museum in Bath. The two girls are dressed rather fetchingly as shepherdesses. Their features are pleasantly strong, although there is a certain wist-fulness in their expressions. Perhaps they had already foreseen their long and genteel confinement to Greys Court. We get a brief sketch of their later lives from the recollections of an old-timer published in the *Henley Standard* on 29 July 1922. When he was young a familiar sight was 'the old Post Chaise, with the red jacketed and booted postilion, which brought the old Misses Stapleton of Greys Court almost daily into Henley'. They evidently kept up appearances.

The preoccupation of the Stapleton sisters with greenhouse horti-
culture was, I dare say correctly, observed by Mr Sharpe. Miss
Stapleton won the first prize at the Henley Horticultural Show in 1837
for 'a boquet of greenhouse flowers'.[16] There are still wooden-framed
greenhouses dating back to Stapleton times within the brick-and-
flint-walled vegetable garden at Greys Court. Catherine Stapleton was
particularly expert on pelargoniums. Her knowledge was recognised
by the honour of having a cultivar named after her in 1826: 'Miss
Stapleton'. It is still available as a variety from specialist nurseries. It
has charming rich red flowers, paler at the base and decked with a
single dark spot on each petal.[17] I have a pot of it on my window ledge.
With her botanical predilections I am certain that Catherine walked
in her own woodland. There she would certainly have found the only
member of her favourite geranium family that grows in Lambridge
Wood (Grim's Dyke Wood included) – the common wayside weed
Geranium robertianum, 'herb Robert'. She, like me, must have bent
down to examine its small, richly red flowers, and must have smelled
its curious pungency, and felt the glandular stickiness of its divided
leaves, so often tinted blood-red, and noted its odd, stilt-like roots.
She too would have known that this herb was named for Nicolas
Robert, a pioneer of accurate botanical illustration in seventeenth-
century France. I can imagine sharing with her a moment's commun-
ion over a mutual enthusiasm before the proprieties of the time sent
her scurrying back to the old house.

Fiddleheads

Ferns have subtle beginnings. As the bluebell leaves fade to little more
than slime, ferns push out their new fronds. In the larger clearing,
fresh shoots of brambles seem to unfold their leaves even as I watch.
Every early shoot – Dylan Thomas's 'green fuse' if ever I have seen one
– is almost soft, and downy, and I have nibbled one and found it pleasant and nutty. Today, the backwardly curved spines lining the veins
on the underside of the newly unfurled leaves are already beginning
to harden – soon they will be capable of delivering a scratch. The

bramble patch is impenetrable and intimidating, and the new growth will serve only to thicken its dense conspiracy. Amidst the scrubbiest part of it are dry, brown, fallen fronds of last year's male ferns (*Dryopteris filix-mas*). From their centre new growth rises assertively. Rebirth started obscurely a month ago as a cluster of dark knobs. Each one soon rears up of its own accord into a fiddlehead, a kind of self-unwinding spiral that uncurls upwards into the spring sunlight. It is rather like that irritating party toy with which children love to blow raspberries at their friends. At the fiddlehead stage it is said to be edible, and I can see a bruised crown where deer have treated the new growth as a seasonal snack. Even now some of the fronds are opening out, like some unfathomable piece of origami, unsheathing the elegant, pinnate blade that will see the year out. The clustered male fern fronds triumph over the brambles. Once the fronds are fully dark green they will be primed with the poisons that have helped them

A 'fiddlehead' male fern.

survive since before the dinosaurs; and then their spore packages will ripen in tiny curved organs beneath each leaflet.

Under drier beech another kind of fern is less difficult to reach, and is more delicate: a buckler fern (*Dryopteris dilatata*), with a triangular frond, finely divided, and broadest at the base. It seems too fragile for such a challenging place where little else grows, and even its fiddle-head is more tentative. The shaft that bears the growing frond is deli-cately clothed in brown, chaff-like flakes. And now on the ground all around this fern so much more brown chaff: little purplish-tan clumps of defunct stamens no bigger than a fingernail have dropped down from the canopy. This is all that remains of the inconspicuous beech flowers. They have already done their job far above me, though the beech leaves are still so new. The greatest trees have the least spectac-ular flowers.

It may seem unlikely that beech leaves could contribute to a deli-cious alcoholic drink, but I have made a liqueur from them for several years, and most of my guests are surprised it is so easy on the tongue. Beech-leaf noyeau can be made in early May when the leaves are freshly unfurled. They are still pale green and soft to the touch – they can be rolled up like cigarette papers. Any tougher and they are bitter. I try to exclude as many of the little brown bracts that originally enclosed the leaf as I can. It takes an unexpectedly long time to pick enough fresh leaves to lightly fill a plastic bag. Once back in the kitchen I stuff a preserving jar quite tightly with the leaves, until it is rather more than half full. Then they are covered with gin (or vodka) until the jar is about three-quarters full. I do not use a high-class brand suffused with many exotic botanicals, but the cheaper stuff from that supermarket shelf marked 'Youths and Alcoholics Only'. I leave the sealed jar for a month to steep. Then the leaves are removed, allowing all the liquor to drain off. If there are any funny bits floating about, now is the time to remove them. For a whole bottle of gin (700 ml) the next ingredients are 200 grams of sugar, around 200 ml of brandy, and 250 ml of water. After boiling the water to dissolve the sugar the resulting syrup is allowed to cool completely. I then add the syrup and the brandy to the beech-leaf elixir and put the mixture back in the preserving jar, preferably with half a vanilla pod. By Christmastide it should be a lovely golden colour. Only a very cynical person would say that it tastes of brandy and vanilla.

Bats!

Claire Andrews has installed her bat monitors. She strapped the recording devices on to our trees about ten feet off the ground, one on the oak by the clearing, the other on a big beech in a sheltered part of the Dingley Dell. They are painted in camouflage colours, and are inconspicuous once in place. They are like discreet garters hitched up on the legs of the trees. Over the next week or so they will record the ultrasonic echolocation noises used by bats to detect their prey, along with their calls one to another.

When I was young I could hear the 'squeaks' of bats, but now I am sadly deaf to such crepuscular cries; yet I have seen dancing, shadowy shapes of bats hunting over our clearing outlined momentarily against a darkening sky, black against indigo. How appropriate is the German word for bat – *Fledermaus*, 'flitter mouse' – which exactly captures these stuttering dashes across the heavens.

It is impossible exactly to identify a species of bat in flight. Our recording machines are attuned to pick up the high-frequency cries of these most elusive mammals. Different species 'squeak' at different frequencies and with different cadences, as they locate and home in on their prey, especially moths. They use echoes to build up a map of their surroundings, rather as the sonar system installed in ocean-going vessels is used to visualise the sea floor. Bats are exquisitely attuned to avoid obstacles in their way, so negotiating a contorted flightpath under our trees poses no problem. Some of their prey species (among them noctuid moths, which are common in the wood) have evolved organs adapted to 'hearing' their approaching nemesis, and will take evasive action if they detect pursuit, such as dropping rapidly downwards from their flight trajectory. Evolution often works as a kind of arms race, with ever more sophisticated methods of attack provoking ever more subtle lines of defence. We need not wonder at the extraordinary auditory organs of the long-eared bats, bizarre though they might appear. These bats 'whisper' with low amplitude and short duration to fool their prey, and they need exceptional hearing from massive ears to detect the tiniest sounds made by insects that they may pick up directly from leaves. By day, all bats hang themselves up like folded umbrellas in secluded roosts. Claire has already spotted several holes in beech trees, and, elsewhere, loose pieces of bark that would afford suitable hideaways. There is nothing to do now except leave the contraptions to do their work.

More than a week later, we feed the digital chips from the recording devices into Claire's computer. Time is ticked off along a chart that reels out on screen the batty history of the glades as night falls. Here is a series of calls from the main clearing at 8.26 p.m. precisely, registering at 45 kilohertz, following sunset seventeen minutes earlier: they appear on the chart as a succession of reverse 'J' shapes, rather like the strokes of an italic pen. 'The one you'd expect,' says Claire. 'Common pipistrelle (*Pipistrellus pipistrellus*).' At 8.39 another batch of short calls appears showing a rather similar shape, but at a different pitch of 55 kilohertz. 'That's the soprano pipistrelle (*Pipistrellus pygmaeus*). It "sings" at a higher frequency.' Claire tells me that the soprano was only named as a species separate from the common pipistrelle in 1999, which seems extraordinary. How could a British mammal elude recognition for so long? We have known all the others for two centuries. Evidently, the two species are extremely similar small brown bats, although they are now known to have different breeding and feeding strategies. As with a lie detector, their voices gave them away. By artificially tuning down the frequencies on the computer we can 'hear' the bat calls for ourselves, and appreciate their different pitches.

At 9.39 a different pattern appears on the screen; it belongs to one of the *Myotis* bats, which are not possible to discriminate on sound alone. Claire believes that our visitor is either the whiskered bat or Brandt's bat, but trapping would be required to say which species. No matter, we will not be following that course. At 10.02 the sopranos return to sing different arias, which show up as sine waves on the screen. These are social calls, aural visiting cards to signal to the group; when rendered into sound I hear repeated chirrups. At 10.12 the distinctive pattern of a noctule bat (*Nyctalis noctula*) appears on the screen; this is one of the largest bats to live in Britain.

Meanwhile in the woodland glade, deep under the beech trees, both types of pipistrelle are dominant, but *Myotis* bats are also flitting through. A distinctive low-amplitude signal identifies the brown long-eared bat (*Plecotus auritus*), and proves that these most delicately adapted hunters are passing under the canopy at 11.16. Claire had expected the long-eared species to appear in this habitat; despite its exotic appearance, it is not rare. This extravagantly outfitted bat may well roost in Lambridge Wood Barn at the edge of Grim's Dyke Wood. The same site would suit a large, and much more uncommon, bat whose signal was identified at 8.40 the following evening: the serotine bat (*Eptesicus serotinus*), a species quite capable of demolishing the big nocturnal beetles that abound under the beeches.

We add them all up. Six different bat species are exploiting the insect life in Grim's Dyke Wood, which must surely be a sign of a generally healthy environment. There may even be a seventh. Claire found one brief signal that might – possibly – have emanated from a snub-nosed, moth-hunting barbastelle (*Barbastella barbastella*), a protected species, and one of Britain's rarest bats. I earnestly wish it to be in our wood, but I know well the emotion naturalists experience as 'the pull of rarity'. It is always so tempting to recognise a more uncommon option. I must rein in my enthusiasm. Until we put up another monitor and get definite evidence from longer calls, the barbastelle bat is 'unproven'.

3

June

Mothing

It is a warm evening when Andrew and Clare Padmore arrive at the wood with their moth traps. Their small generator powers a bright light set in the middle of a stage. Beneath this platform the moths that are attracted to the light can tumble down into a container full of *papier-mâché* eggboxes. The light goes on at dusk and we sit under the beech trees on the edge of the large clearing waiting for darkness. Somewhere further away in the wood there is a noise made by some moderately large animal passing through; it is probably a badger somewhere near Grim's Dyke. The night embraces us. The artificially illuminated beech trunks fade away a little spookily in the distance into far blackness.

The first moth – a beautiful Green Carpet Moth (*Colostygia pectinataria*) – comes out of the dark and desperately flutters around. It flops on to the ground sheet, and then off and around again until trapped in a jar where we can admire its triangular form and chequered green markings. As if from nowhere a big, hairy moth arrives. It has pale, furry legs which point forward as it rests, and exquisite, comb-like, brown antennae – Andrew identifies a Pale Tussock Moth (*Calliteara pudibunda*). It sits very still as if bemused, hind wings tucked under the forewings, which are marked with an impossibly complex, undulating greyish mottling. This particular species does not feed as an adult; its job is simply reproduction. Then comes a smaller, darker species, the Nut Tree Tussock (*Colocasia coryli*). 'They are all,' says Andrew, 'in the peak of condition, just emerged from the pupa.'

Feathered antennae distinguish most moths from butterflies, which have comparatively slender ones carrying knobs at the tips, and it is clear that our moths' antennae are working away even now, twitching and sweeping. They are hypersensitive chemical sampling kits smelling out messages borne on the night air: odours from freshly unfurled leaves as food for their caterpillars, or the attractive pheromones that identify their mates. Theirs is an olfactory world; light is almost superfluous. I have a vision of the night air as a miasma, dense with molecular messages that only moths can read. They do however use the moon for navigation – our lights serve to confuse their direction-finding, which is why the insects arrive in our collecting boxes.

They are not alone: two fat, succulent cockchafer beetles – May bugs (*Melontha melontha*) – prove that other creatures are also abroad. The big brown beetles scrabble at the light, looking oddly like cockroaches with ill-fitting wings. There is something repellent about their insistence. Although their larvae cause damage to plant roots the leaf-eating adults are harmless enough.

Now my eyes are fully accustomed to the darkness. The sky is visible in places between the interwoven crowns of the trees. It is not as profoundly dark as the distant recesses of the wood; it is rather an ineffably deep blue dotted with stars. As I look upwards the lamplight catches on horizontally disposed beech branches, making drapes of them, a series of stacked canopies fading upwards. Our sampling site has become a kind of theatre, with beech trunks making the proscenium columns, framed by swags of real leaves. Two small bats now flutter into the auditorium, briefly picked out by the illumination: in and out, and then again. Will they scoff the moths we have worked so hard to attract? When a Brimstone Moth (*Opisthograptis luteolata*) arrives even I, a moth beginner, can identify it, since apart from a few reddish splashes on the front of the wings it is all brilliant sulphur yellow. In contrast, the Waved Umber Moth (*Menophra abruptaria*), the size of a small leaf, is so perfectly disguised it looks like a fragment of animated tree bark; at rest during the day it is invisible. New arrivals continue. The light attracts a kind of living fuzz of many other tiny insects I cannot identify. They all have secret livings to be made in the wood, if only I could know what they were. Somewhere in the distance a screech owl cries, but not so fiercely, as if in sympathy.

Andrew Padmore will return to the wood many times. More and more moth species will be attracted to his lure, which is later replaced by a solar-charged model hidden deep in the trees. No harm is done to the gentle moths: a photograph is taken and they are released to go about their business. As I write the list of species recovered has now climbed beyond 150. Different moths are on the wing at different seasons, finishing perhaps with the November Moth. There is a curious poetry about moth names, which is an esoteric language of analogy, allusion and colour. The wood has yielded more than half a dozen different species of carpet moths. There are several pugs and rustics, thorns and swifts, footmen and oak beauties. Who can resist the Chinese Character, the Coxcomb Prominent, or the Feathered Gothic? Or Bloomer's Rivulet, the Rustic Shoulder Knot, Blood Vein and Mocha? They are all in the wood. Sometimes the common name is a simple description: the Blood Vein does indeed have a single, bloodily tinted vein describing a clear line like a gash across the middle of the wings. The Chinese Character does carry a distinctive pictogram; but it more closely resembles a bird-dropping when at rest. The Flounced Rustic is a furry, wonderfully complex, mottled and blotched mass of tans and greys; but I fail to see the flouncing. The Mocha is a nationally scarce buff-and-brown moth that maybe suggested coffee to some entomologist in the early days of the science. All the names have charm. Nobody could argue about the origin of Peach Blossom (*Thyatira batis*); it is marked as if some evolutionary leprechaun had implanted a few whole, pink flower heads on the darker forewings – just for fun.

We caught some moth species only once; they probably included wanderers from grasslands and gardens, feeding on plants that are not found in the wood. I would have loved to find more hawk moths, but we don't have poplars or convolvulus to nourish their caterpillars. The moths most commonly trapped are naturally those whose food plants are present in Lambridge Wood. They are an intrinsic part of the ecology. The incomparable Peach Blossom is a bramble feeder, our commonest shrub. The most abundant species of all was trapped 111 times: the Clouded Magpie (*Abraxas sylvata*), a large and very pretty white moth blotched with patches of orange-brown, grey and black. Its food plant is wych elm, and Grim's Dyke Wood has plenty of wych elms. Andrew had never realised that it could be so numerous – but then, elms are not so widespread these days. The Gold Swift (*Phymatopus hecta*) is one of the few insects that can feed on bracken, that *potpourri* of pernicious poisons, and does not have far to fly to find its favoured larval foodstuff. The little brown Snout Moth (*Hypena proboscidalis*), all pointy at the front and the shape of a tiny delta-wing aeroplane, needs nothing more than nettles. Despite its name, the Willow Beauty (*Peribatodes rhomboidaria*) can feed on tough ivy. This moth is a wonderful confection of brown and black speckles on a buff background – the very embodiment of the word 'cryptic'. It is so cryptically coloured the wonder is that the lepidopterists ever discovered it at all. The Satin Beauty (*Deileptenia ribeata*) is almost as well-disguised, and can feed on uncompromising yew needles. Then I must catalogue forty Lobster Moths (*Stauropus fagi*), dullish-coloured and almost as big as your thumb, and very plump and hirsute; as their Latin name implies they favour *Fagus*, and there are beech trees as far as the eye can see.

The Lobster Moth reminds me of an interesting puzzle. In spite of the wealth of its lepidopteran life I have noticed very few caterpillars since I have owned the wood. I have to conclude that this 'eating machine' stage of the moth's life takes special trouble *not to be observed*: a green body on green foliage, stick-like mimicry, rolling a leaf into a private self-service restaurant – these are some of the tricks of the larval trade that different species employ to avoid a questing beak. Only very poisonous species like to announce themselves in yellow and black stripes. On a hazel stick I did find the caterpillar of a member of the geometer family (it might even have been that of the Brimstone Moth), a typical 'inchworm' with legs only fore and aft along the body, so it progresses by looping up its midriff as it brings its hind legs forward. Measured steps are not an inaccurate description (hence the geometry). When it stops under the threat of my close eye it raises one end into the air and becomes a twig. Even more, it shows countershading. That is, its upper part is darker than its underside. Normally, things lit from above are relatively illumined on that side, which makes them more conspicuous. By introducing compensating darker tones on the dorsal part of the body such contrasts are flattened out: the object (well, inchworm) melts into the background. As they say on soap powder advertisements: it really works!

As for the Lobster Moth, high in our beech canopy, it is a deceiver to dumbfound John le Carré. When the larva first hatches from the egg it is an ant imitator, with spindly legs that wave around a lot, and it thrashes about like an injured ant if it is disturbed. The young caterpillars are reported to defend their egg territory, and will drive off any rival caterpillar that comes too close. As they moult and grow they become both voracious leaf consumers and very odd looking – one of nature's gargoyles. The head is larger and the legs behind it (the thoracic legs of the adult) become unnaturally attenuated even as the four pairs of legs further behind become stumpy and grasping. The back gets covered in humps, and the tail end can turn back on itself like some kind of turgid bladder, all finished off with a spike. The entire caterpillars develop a shade of pinky brown, and since they can be seventy millimetres long fully grown they are quite enough to give a shock to any casual stroller who comes across one; especially when their body is raised in the threat position with the head arched back. It is said to resemble a cooked lobster; it is certainly scary.

I wonder if all of our 150 or so moths have such complex tales to tell. The beech canopy is humming with life stories, the brambles alive with deceptions and role-playing, each crack in the bark of every tree a dark dive hiding darker narratives.

Beech

By June, the beech canopy has garnered all the light, each leaf second-guessing its neighbour at grasping any space giving on to the sky. The taller trees soar upwards for more than a hundred feet. From the ground they seem all trunk, but from the sky they seem all crown. The beech (*Fagus sylvatica*) has always been a working tree: for furniture, fire and faggots. John Evelyn's *Sylva*, the first book published by the Royal Society in 1664, and the founding text of forestry, said of beech trees: 'they will grow to a stupendous procerity, though the soil be stony and very barren: Also upon the declivities, sides, and tops of high hills, and chalky mountains especially.' Evelyn then quotes an old rhyme:

> Beech made their chests, their beds and the joyn'd-stools,
> Beech made the board, the platters, and the bowls.

Three hundred years ago, beech may not have built the houses, but it did almost everything else. The management of beech trees has been the story of our wood for centuries.

In 1748, Peter Kalm, a Finnish protégé of the great Swedish botanist Linnaeus (who named the beech tree scientifically), made an informed journey through the woodlands of England.[1] He observed the Chiltern lands at Little Gaddesden, a short distance from our wood over the Buckinghamshire border. Some of the trees he saw might indeed have been our own, for 'the beeches are for many fathoms in their lower part entirely without branches, and quite smooth'. The woodsmen climbed the trees in search of squirrels (at that time red squirrels), or rooks' nests to provide the table with squabs. They rarely used ladders; instead they strapped hideously sharp 'crampoons' to their feet to scale the trees, like some oversized squirrel themselves.

Kalm recorded precisely how, after felling, every part of the tree had a value; almost nothing went to waste. Farmers used to say of pigs that everything is used except the squeak; the beech woodsmen's equivalent might be: everything has a use except the bark. They 'sold the smooth part, or sawn it up into boards, but those of which the stem had been knotty or uneven was cut up for firewood and piled up in cords. When the beeches … were cut down and felled to the ground they were cut off close to the earth. Two or three years after that, the stump that had been left, together with all the roots proceeding from it … was dug up, cut into small pieces and arranged in four sided oblong heaps to dry … In digging up the roots they had been so careful that among those heaps there lay a great many fibres of the roots, whose length was not over 6 inches, and thickness not greater than a quill pen. These roots thus arranged were sold as fuel to those who lived some English miles around.' Dry twigs bound into bundles of faggots were fuel for bread ovens. Some observers even regarded the beeches in the way that we now look at factory farming. The pioneering landscape architect Humphrey Repton remarked in 1803 that 'these woods are evidently considered rather as objects of profit than of picturesque beauty'. He preferred specimen trees carrying full crowns of branches adorning a grand park, the whole designed for effect. He would not have stooped to grub up roots.

Kalm also made calculations, and his observations show a clear, scientific mind at work. 'A beech trunk was measured which had at the large end fifty four sap rings. The diameter was just two feet. The sap rings which were found nearest the heart were narrowest and smallest, from which they grew larger the further they lay from the heart out towards the surface.' A cross-section cut from the trunk of a tree could not have been better described. The 'sap rings' are the record of the new wood lain down by each year's growth beneath the bark: fifty-four rings is fifty-four years, the age of the tree. Our own wood needs just such a chronology.

The neighbouring wood has had some recent felling, and I can record the cleanly cut log-ends on display in a stack by the entrance to Grim's Dyke Wood. Beech chronology turns out to be not quite as simple to measure as I might have thought. The good thing about our trees is that such straight trunks provide a reliable, nearly circular section. Nearer the ground the trunks are all buttressed and irregular, and no two diameters are the same; these undulations record the profiles of the 'props' that hold the trunks aloft. So the upper part of the tree – waist-height and above – provides the best experiment. Since all tree trunks do taper gently, different sections of the same tree will have decreasing diameter upwards. The difficulty is that the 'narrowest and smallest' rings in the centre of the tree are not so easy to read. Some years added no more than a millimetre of new wood.

Stacked beech trunks displaying growth rings.

I have to take a felled piece of heartwood home to see if I can tease out some figures. I laboriously buff it with fine sandpaper for hours, and as the distracting irregularities are polished away, so the early growth rings become clearer as darker lines. It is like seeing a diagnostic thumbprint slowly developing from obscurity. The wood almost shines pink-brown when I finally make out twenty-seven rings in thirty-five millimetres diameter. It evidently took a long time for this particular tree to get going, after which it sped up mightily. Even in the mature part of the tree not every ring announces itself clearly. There are good years and bad: the summers of 1974 and 1975 were droughts, and the growth rings would have been minimal. Skilled dendrochronologists can 'read' tree rings as a diary of climatic variation extending over centuries, but my skills do not extend that far. However, in older trees most of the rings add about three to four millimetres to the radius every year, and these can be counted easily enough. I eventually reach a consensus with my own scientific conscience. Several trees come out with eighty rings, more or less, possibly as many as eighty-five. Jackie provides a second pair of unbiased and independent eyes and tots up a similar figure. These are from trunks ranging in diameter from twenty-seven to fifty centimetres; and another trunk of forty-three-centimetre diameter has just under sixty rings. I cannot prove that the former come from higher in a tree that might have had a more impressive base. What I can say, with confidence, is that a number of beeches in Lambridge Wood grew from seedlings around 1930, and are now fine, big trees.

It is easy enough to convert diameters into circumferences, and with my very own trees the latter is what I record at shoulder-height with my tape measure. I can prove that many of the standing beeches are of similar size to those sitting on the log pile. It is actually rather easy to show this *without* wielding the tape, by using that alternative, hippy measurement – 'the hug'. Trees with a fifty-centimetre diameter can be comfortably hugged, with hands meeting around their girth. There are an equal number of trees that are just too big to hug, although they do decrease in diameter to become huggable towards the canopy. And then there are the real giant trees, like the King Tree and the Queen Tree, and one I call the Elephant, with circumferences up to 250 centimetres. Surely these are much older than eighty years. If I assume that they continue to grow by adding a three-to-four-millimetre ring every year, it is not unreasonable to arrive at an age of 140 to 180 years. There are perhaps a dozen of these trees scattered through our wood. Their bark eventually loses the smoothness of the younger trees to become lightly scarred, as if daubed with vertical stretch-marks. Since there are certainly no trees still more antique, I conclude that these fine examples have been responsible for seeding some of their younger companions. They have been left alone. A great

felling must have occurred about eighty years ago – and selective felling probably continued for another twenty years or so until Sir
Thomas Barlow's ownership, when we know that little happened in
our part of the wood. The somewhat 'unhuggables' may well record
regrowth after another, earlier phase of beech harvesting. There is no
doubt at all that the whole wood has been replaced, thinned, sawn and
regenerated. Its history is written in the tree rings. This is the same
wood that John Stuart Mill walked through in 1828. Only the trees
have changed.

Like those of many wind-pollinated species, beech flowers are
unspectacular. I already noticed brown bunches of fallen stamens
from the male flowers in May, while the separate female components
now sit above, waiting to mature into three-sided beechnuts, which
will eventually fall to the ground in October. The most beautiful and
accurate drawings I know of living twigs are by Sarah Simblet in *The
New Sylva*,[2] which is a large, luxurious, even sumptuous tribute to
John Evelyn's original, and about as appropriate for taking into a real
wood as *The Oxford English Dictionary*. Last year's beechnuts germinate as early as April, and the seedlings can be told from all others by
their pale-green seed leaves (cotyledons), which resemble the blades
of two inch-wide ping-pong bats placed side by side. Before the
canopy has opened out, optimistic seedlings can come up almost
anywhere in the beech litter, and are not short of light. A tender shoot
then appears between the two seed leaves and starts to put out regular
leaves. By now in June it is already clear that most of these young
plants are doomed; they lack enough light to make further progress,
as the canopy sucks it all up to feed the crowns of the trees. The babies
yellow and fade. Only those seedlings close to a clearing can put on
the vital first inches of growth that will give them a chance to mature
into a giant. That is where a dozen or so small beech trees not much
taller than I am vie to be first to fill the gap in the sky. At some stage I
will have to pick a winner and thin out the rest. If I fail to do so the
surviving trees will become too crowded and grow all spindly.

Squirrels

I am sitting in contemplative mood on a beech log left behind by cousin John when the bombardment begins. I cannot work it out at first. Bits of hard stuff are falling from the sky, and some of them are hitting me. Then I catch a piece as it lands: it's a fragment of beech bark, more than a quarter of an inch thick. I am being pelted with beech bark! Protecting my eyes with spread fingers I look for the source of the onslaught. Perhaps forty feet above me a horizontal beech branch leans out from the nearest trunk. I catch a glimpse of something grey and fuzzy moving about on top of the branch. Then a squirrel peeps momentarily over the edge and identifies itself; it is not worried for its safety – it is only concerned about lunch. It is obviously not eating the beech bark; it is throwing it at me instead. He is after the sugary spring sap still flowing beneath the bark. Like one of the regulars in the Maltsters Arms, he is having a liquid lunch. The bark is stripped and the layer underneath it licked clean. It is obviously damaging to the tree. Now I notice that the bole of a nearby beech – and not a small one, either – displays a raw wound. A patch of bark has been removed, and the sapwood is on display, all yellow and unnaturally bright. Several other trees around me show the same feature, always close to the roots. In my absence, the squirrels have been picnicking *al fresco*.

This arboreal dining habit explains a feature I have noticed on fallen beech branches. Many of them have the bark stripped from the upper side; this is less obvious than on new wounds because the colour contrast has dulled with the passage of time. Bark on the undersides of the branches is protected from squirrel activity, so seen from the ground branches high above look just fine. In fact, many are damaged on top, and perhaps this encourages them to fall before their time.

Another chip of bark whizzes past my ear. I could almost hear a snicker from far above. Re-examining the chewed boles of the beech trees I see yet more evidence of old scars. Fortunately there is enough bark left to allow the big trees to survive. Nor is all well with some of

Grey squirrel damage to the base of a fine beech tree.

my young beeches. Many of those with trunks thicker than my arm have been mutilated in a similar way. A few trees of middling size – forty years old perhaps – have become grotesquely distorted, their crowns twisting like corkscrews, branches all whiskery and set akimbo like broken limbs. I had not known what to make of them before. Squirrel damage has stunted and deformed them. 'Little bastards,' I growl, but that hardly seems adequate for an animal that may be affecting beech regeneration that has hitherto endured in the Chiltern Hills for a thousand years.

There are always grey squirrels somewhere in the wood. They skitter acrobatically along branches and leap effortlessly through the canopy; it is their realm. They build untidy drays high in the trees in which they can raise two litters a year. They have abundantly fluffy tails. They are, of course, invaders from North America. They were released on a few English estates in the nineteenth century for aesthetic reasons, and then stayed on and prospered. They pushed out the red squirrels from most of England, and continue to expand their range northwards into Scotland today: they are bolder animals, faster breeders and generally more robust. They carry a lethal pox virus to which their red cousin has not yet acquired immunity. There is nothing new about worrying about the invader. A wartime *Surrey Mirror* exclaimed in 1942 that to eradicate this pest 'all possible steps such as shooting and trapping must be taken. The national interest demands it.' Never mind Hitler: the nation might be brought down by a climbing rodent! When I was a youngster there was a bounty of sixpence on every grey-squirrel tail. Neither threats nor inducements have worked: the cheeky grey squirrel dances nimbly onwards.

It has been claimed that red squirrels are better adapted to conifer woods and that greys outcompete them only elsewhere – though I know plenty of conifer plantations with greys in command. I try very hard to banish Beatrix Potter's charming *Tale of Squirrel Nutkin* from my mind, since her drawings provide such effective propaganda for the red species. Some ecologists even challenge the notion of 'native' species at all, when so much British wildlife has come from elsewhere. They are probably right that it is foolish to think of restoring some notional Eden, a prelapsarian paradise labelled 'Natives Only'. In this argument I am obliged to take the part of my precious beech trees. Although I can find some records of tree damage by red squirrels, it does not seem to be as extensive as that caused by the grey interloper. A proven continuity of fine beech woods in our patch points to the red squirrel as no more than an occasional nuisance. Maybe they were once popular enough as food to keep the numbers down. Most damage happens in years when lots of squirrels have come through a mild winter following a good year for beech mast: overpopulation is part of the problem. One of my woody neighbours shoots as many greys as he can; another believes nature will correct the numbers in her own good time.

I have found a bleached squirrel skull to add to the collection, manner of death uncertain. I simply want to believe that there will still be tall, healthy beech woods here in the century to come, so that some future J.S. Mill may glory in their abundance. In the end, global warming might be more important than any kind of squirrel. *The New Sylva* warns that if summer drought increases, beech 'may disappear from the Chiltern Hills except on northern slopes with moist soils'. To survive at all, the woods will have to move northwards, alongside the delinquent greys. They will become partners in a human crime. I shudder at the thought.

Two ghosts and a Dutchman's pipe

When the beech canopy captures the sun the forest floor becomes a darker place. The bluebells have faded, and only faint greasy traces reveal the wraiths of their dead leaves. The grasses that made a brief, bright sward are muted now; nodding wood melick has set its seed for the year and will soon aspire to invisibility. Taller, elegant wood millet (*Milium effusum*) raises its flowering spike in wispy tiers making a brief show of green flowers that dangle from the ends of spread branchlets like tiny beads. Only sedges by the wayside are more obdurate. Their tufts and clumps of coarse, dark-green leaves see out the seasons, though few passers-by would notice them if there was the bright promise of bluebells in the woods beyond. Wood sedge (*Carex sylvatica*) briefly dangles little rods dotted with yellow stamens in spring, and then might even be described as pretty.

Its broader-leaved companion, thin-spiked wood sedge (*Carex strigosa*), is a *recherché* plant for botanical enthusiasts, with flower spikes so discreet that I can only identify the species with a lens in one hand and a book in the other. It is something of a rarity, though it seems to grow enthusiastically enough in ruts left by tractors. Distant sedge (*Carex remota*), with the thinnest leaves of all our sedges, lurks inconspicuously by the damp seep, and has its greenish fruits tucked into its leaf bases, so that anyone giving it a casual glance would think it a grass. Toughness in sedges is evidently inversely proportional to their showiness. But even they cannot grow under the largest beeches. Apparently, nothing can. The deepest leaf litter is inimical to living things. It is a place fit only for ghosts.

Nina Krauzewicz's sensitive drawing of the
three sedge species in the wood.

The rarest plant in Britain is such a ghost: the ghost orchid (*Epipogium aphyllum*).[3] It disappears like a phantom and then conjures a new haunting in a new wood. It has been declared extinct, and then spookily reappeared, even after decades. It is a spectre much sought by botanists; some plant-hunters develop an obsession with its rediscovery. And it made one of its few and unexpected appearances in Lambridge Wood. Ninety years ago a young Henley woman called Eileen Holly found it in deep litter where nothing else will grow. It appeared from 1923 until 1926. A lively eyewitness account from a prolific botanical diarist, Eleanor Vachell, leaves no doubt about the drama of the discovery – even though it reports the ghost of a ghost:

28 May 1926. The telephone bell summoned Mr. [Francis] Druce to receive a message from Mr. Wilmott of the British Museum. *Epipogium aphyllum* had been found in Oxfordshire by a young girl and had been shown to Dr. [George Claridge] Druce and Mrs. Wedgwood. Now Mr. Wilmott had found out the name of the wood and was ready to give all information!!! Excitement knew no bounds. Mr. Druce rang up Elsie Knowling inviting her to join the search and a taxi was hurriedly summoned to take E.V. [Eleanor Vachell] and Mr. Druce to the British Museum to collect the particulars from Mr. Wilmott. The little party walked to the wood where the single specimen had been found and searched diligently that part of the wood marked in the map lent by Mr. Wilmott but without success, though they spread out widely in both directions ... Completely baffled, the trio, at E.V.'s suggestion, returned to the town to search for the finder. After many enquiries had been made they were directed to a nice house, the home of Mrs. I., who was fortunately in when they called. E.V. acted spokesman. Mrs. I. was most kind and after giving them a small sketch of the flower told them the name of the street where the girl who had found it lived. Off they started once more. The girl too was at home and there in a vase was another flower of *Epipogium*! In vain did Mr. Druce plead with her to part with it but she was adamant! Before long however she had promised to show the place to which she had lead [sic] Dr. Druce and Mrs. Wedgwood and from which the two specimens had been gathered. Off again. This time straight to the right place, but there was nothing to be seen of *Epipogium*!

2 June 1926. A day to spare! Why not have one more hunt for *Epipogium*? Arriving at the wood, E.V. crept stealthily to the exact spot from which the specimen had been taken and kneeling down carefully, with their fingers they removed a little soil, exposing the stem of the orchid, to which were attached tiny tuberous rootlets! Undoubtedly the stem of Dr. Druce's specimen! Making careful measurements for Mr. Druce, they replaced the earth, covered the tiny hole with twigs and leaf-mould and fled home triumphant, possessed of a secret that they were forbidden to share with anyone except Mr. Druce and Mr. Wilmott.[4]

It is a measure of the allure of this botanical will o' the wisp that even a cut stem provoked such delight. The flower in person might have induced a serious attack of the ghostly vapours. It is a pretty enough plant, with a few, quite large blooms for a European orchid, each with a pleasingly pinkish spur and yellower sepals. It is fragrant, and probably insect-pollinated. But the plant has no leaves. It has no green on it anywhere. It consists only of a flower spike and the 'tuberous rootlets', or 'coral-like rhizome', as V.S. Summerhayes described it in *Wild Orchids of Britain*[5] (this led to an alternative common name of 'spurred coral root'). The scientific species name *aphyllum* even means 'without leaves'. Since the flowers blend almost perfectly with beech leaves as a backdrop it is little wonder that they so readily escape detection: it's a ghost in camouflage. It seems to be an impossible plant, because it has no chlorophyll to manufacture vital proteins and sugars. It clearly does not need light; it can grow in deepest shade where no other plant flourishes. In my old copy of Summerhayes, the author attempted to solve the mystery by allowing the ghost orchid to get its nutrients 'already manufactured' from 'the humus of the soil, which consists of numerous more or less decayed parts of plants and also animals'; in other words, to grow like many fungi – which never have chlorophyll. The story is much more nuanced than that, although mushrooms do indeed play a part, as we shall see.

After Eleanor Vachell's visit the orchid vanished from Lambridge Wood. Stirring up its rhizome would not have helped. Joanna Cary, who lived nearby and was wont to wander in Lambridge as a child, tells me that in the 1950s she used to avoid crossing paths with funny men in gaiters up in the deep woods, and assumed they were flashers, or worse still, burying something unspecified. It was probably Mr Summerhayes and his eminently respectable band of ghost-hunters. Another local plant enthusiast, Vera Paul, continued the botanical tradition by finding *Epipogium* at a site just a couple of miles away, sporadically, for over thirty years until 1963. I have seen a drawing of the famous plant framed on the wall at her former house in Gallowstree Common. More recently, the orchid disappeared completely for more than twenty years, until a remarkably persistent ghost-pursuer, Mr Jannink, rediscovered a small example in 2009 in one of its old sites near the Welsh border, far, far away from the Chiltern Hills.[6] The ghost orchid is not extinct in Britain after all. However, nothing I have read explains how a plant with such minute seeds can apparently jump so dramatically from place to place. There is something almost spooky about it.

Another ghost haunts Lambridge Wood. Nobody has actually seen it, but I am assured its presence has been *felt*. After dodging the 'dodgy' gentlemen, Joanna Cary also avoided 'the murder cottage'. Nowadays, it is a pretty house adjacent to the barn at the very edge of our wood, but its reputation must have lingered on for decades. As the *Henley Standard* reported at the time: 'Friday, December 8th 1893 will always be regarded as a black day in the annals of Henley history'. The body of the thirty-year-old housekeeper who looked after the farmhouse, Miss Kate Dungey, was found in the woods a few yards from the door with 'a terrible gash in the left side of the neck, and a number of wounds about the head'.

It was quite the shock headline of the day; the gruesome story was reported prominently as far away as New Zealand. It had all the right ingredients to impress the public. 'The spot is as remote and lonely as could possibly be found, and there is very little likelihood of cries for help being heard,' the *Standard* reported; and naturally 'it was a dark and miserable evening'. Miss Dungey was an interesting victim, 'of good figure, had dark hair, and is said to have been good looking' – moreover, she was an ex-governess for the children of Mr Mash, fruit-erer and owner of the house, so she had the trappings of a gentlewoman. 'Almost all around Henley knew Miss Dungey and speak well of her,' the newspaper continued. Could robbery be a motive when 'nothing had been touched in the house, not even the watch on the sitting room chair'? There were signs of a struggle and blood by the front door, so perhaps the grisly killing took place as the poor woman attempted to flee her assailant. A thick, cherry-wood cudgel discovered by the body may have been involved, but something much sharper caused the deep gash.

Over the next month new evidence emerged, as well as rumours that Miss Dungey had a romantic interest in a local married man, details of which never appeared. By 3 January 1894 one Walter Rathall had been arrested for the crime. He had worked as a labourer on the farm, and led a rackety and irregular life, being at times little more than a tramp. *Jackson's Oxford Journal* reported on 13 January that Rathall slept out in the woods all the previous summer – *our* woods. The paper described how Kate Dungey had advanced him money, which she never recovered, and that 'he had been discharged principally through the instrumentality of Miss Dungey, with whom he had several quarrels'. Despite an apparent motive, the circumstantial evidence gathered by the police proved insufficient to secure Rathall's conviction. He walked free; the murder mystery remained unresolved, as it still is to this day.

Hayden Jones, the current occupant of 'the murder cottage', tells me a ghost story. On the hundredth anniversary of the murder it was another 'dark and miserable evening', though cosy enough inside the house. Hayden relates that the company decided to have a toast to the memory of 'poor Kate'. As the glasses were raised all the lights in the house were suddenly extinguished – poof! Hayden had previously encountered a definite reluctance on the part of certain woodsmen to enter his premises: a shake of the head and a polite refusal. A presence, they said. It is all nonsense, of course, as every rationalist will agree. Yet, since I heard the story of Miss Dungey, I have been in the wood on an overcast, windy evening late in the year when I heard a sudden brief, distant cry – it must have been a red kite out late, or even a frightened blackbird. And a crunching noise behind the holly bushes was surely just a small, squirrel-weakened branch falling suddenly and noisily to the ground; it is no restless murderer's shade on the march. Ignore the sudden shiver. Let's not be silly.

*　*　*

The tortuous saga of the ghost orchid prompts me to make a thorough quartering of Grim's Dyke Wood in June. It is too much to ask of my tiny piece of ground, I know, but that does not stop me peering closely at every beech-leaf-filled gulley. I will not miss a thing, I tell myself, and for half an hour I trudge like a botanising zombie up and down, up and down. For an instant, my heart stops. Here are two yellow stems arising from the ground and bearing flowers. There is no sign of a leaf, or anything green. So is it an orchid? The stems curve over at their apices like shepherds' crooks where perhaps half a dozen yellow flowers hang down, almost in the fashion of our bluebells; however, these flowers are tubular. This is not a shape known from any orchid. This may be no ghost, but it still thrills like a sudden, strange apparition. The *Red Data List*[7] records some of the most precious and uncommon species of plants in Great Britain, and this is one of them, in our very own wood! I have known it for many years as an illustration in the Reverend Keble Martin's indispensable *New Concise British Flora*. In an even older book I have a list of all the wildflowers I have ever seen, which I have been ticking off since I was a boy: this is one plant that had remained persistently unticked. Nor is it some anonymous, tiny green herb. It is another special plant in the ghost orchid mould lacking all chlorophyll, a spooky spectre, and somehow implausible. It is called the Dutchman's pipe, or if you prefer, yellow bird's nest, and by scientists *Monotropa hypopitys*. I have never met a pipe-smoking Dutchman, but I would now recognise the shape of his favourite accoutrement.

On my hands and knees, I brush away a few loose leaves concealing the bases of the stems of the new discovery. They look a little like blanched asparagus spears, complete with scattered scales. They are the only plants growing in the deep shade. They really do rise straight out of the ground. I would be willing to bet a hundred squirrel tails that if I dug down they would originate from swollen roots such as Eleanor Vachell found for *Epipogium*. I am not going to try it. A small beetle emerges from one of the flowers, having, I suppose, helped to fertilise it. Over the next few weeks I keep tabs on the small blooms: they last and last. The Dutchman's pipe is not taking many risks when it comes to setting seed.

Monotropa has recently been the focus of botanical research. In my old edition of Keble Martin – and in many later books – it sits all by itself in its own plant family (Monotropaceae). It seems that no expert could quite make up his or her mind where such a weird, penumbral paradox fitted into the grand scheme of plant evolution. In North America a related, almost supernaturally pallid species is known as the Indian, rather than Dutchman's, pipe, or sometimes as 'the corpse plant' (*Monotropa uniflora*), which suggests that we are never going to be able to escape the whiff of the graveyard in this chapter. When the techniques of molecular analysis to determine ancestry became widely available it was not long before both species of *Monotropa* were allied with a much larger plant group, the Ericaceae, the familiar heather (or blueberry) family, with something like four thousand species worldwide. The Dutchman's pipe was, in its essentials, a heather that had lost everything above ground except the flowers. Now that I study them again, the flowers of *Monotropa* do indeed recall those of strawberry trees, blueberries or bell heathers – perhaps we should have known all along. Occasionally, science just reinforces common sense.

The root of the ghost puzzle really *is* the root. All our ghostly plants, whether orchid or pipe, have similar-looking roots, which are tuberous and puffy. Both the loss of chlorophyll and the ability to thrive under the beech canopy are the result of special adaptations secretly hidden away underground. V.S. Summerhayes was right in essence: neither the Dutchman's pipe nor the ghost orchid manufactures its own nutrients. But he was wrong to assume that these plants were what he termed 'saprophytes' – that they sourced all they needed from the rotting leaf litter surrounding them. The explanation is both more complicated and much more wonderful than mere scavenging. *Monotropa* and *Epipogium* are playing parasitic piggyback on mushrooms. In the case of the Dutchman's pipe the fungus has been identified with an ordinary-looking mushroom that has been called the girdled knight (*Tricholoma cingulatum*)[8] – not exactly a regular 'shop mushroom', since it has a greyish cap and white gills, but constructed along the same familiar lines. Our pallid plant has given up any attempt to manufacture its own necessities in favour of stealing all it wants from its fungus host. Above ground, it needs to be nothing more than flowers and seeds. Like some Regency dandy feeding off colonial slavery, the organism can be all show and no hard graft. The distinctive roots of the plant reveal the truth: they are full of fungus, and modern techniques of DNA analysis allow the molecular biologist to identify exactly which species from a choice of thousands. When I started out in science as a botanising youth this would have been impossible, but now it is almost routine procedure back in the laboratory.

However, this is not the end of the story. For the fungus itself lives in an intimate association with beech trees in deep woodland. The 'roots' of the fungus are masses of threads called mycelium. These threads move through the moist soil seeking out nutrients, and they are skilled in reprocessing all that mush and drift of rotting leaves. Mycelium is the workhorse of the fungus, while the familiar mushroom fruit body is just the culmination of the life cycle for spreading the minute spores of the species. Like many other fungi, *Tricholoma* forms a partnership with the roots of beeches, where it can live for many years. The threads of mycelium fully coat the growing tips of the roots rather as tight-fitting kid gloves enclose the fingers, and the fungal talent for acquiring important foodstuffs such as phosphates from the surrounding environment becomes essential for the healthy growth of the tree. The fungus-coated rootlets seek out valuable molecules. The fungal dressing is called mycorrhiza, which is simply a classical way of saying 'fungus root'. Mycorrhiza makes for a reciprocal partnership, because the tree in its turn does what it does best – manufacturing sugars and other products of photosynthesis – and supplies them to the growing fungus, which cannot make them for itself. It is a *symbiosis*, an intimate growing-together. Like a well-honed comedy duo, each partner would fall flat without the other.

So the Dutchman's pipe is at the foppish apex of a *ménage à trois*. The beech works with sunshine and rainfall, and supplies the fungal partner on its roots with the means to quest for more exotic vital nourishment. *Monotropa* is a parasite on the fungus, so indirectly it too benefits from the photosynthetic work of the lofty beeches, and can dispense with its own green parts. The fungus supplies everything else. Freed from the need for light, the parasite can safely flower in deeply shady glades where nothing else can prosper.[9] Nor does it have to flower every year. In a bad year for either tree or fungus it can hang on as a root or rhizome hidden beneath the litter, biding its time. Now we can understand the fickleness of those ghostly appearances. The spooks might really be there all the time.

Cherry-picker

A cherry-picker comes to the wood to access the canopy. Shane, the operator, a young man with dramatically pierced ears, has brought it all the way from Essex on the back of a truck; it is a special piece of kit. The contraption trundles through the wood on caterpillar tracks until it reaches a place where it is possible to shoot upwards through the trees. Then four legs like those of a spider with long flat feet extend outwards on to the firm ground to support a lifting platform. I am first in the queue. A harness is strapped to me, and I step gingerly with Shane on to the small, railed platform.

Although they look quite neat, the telescoped shafts of the machine are capable of extending to more than ninety feet. The platform twists and swivels, guided by Shane's handset, as it rises to avoid overhanging beech branches. The ascent is like passing through a series of extended curtains decked in fresh leaves. We rise speedily, the foliage occasionally brushing my face, and then – quite suddenly – we break through the canopy. We must be at least eighty feet off the ground. I am too fascinated to feel at all scared. The treetops billow out in all directions, an extraordinarily rough sea of breaking waves of foliage, and above, nothing but the sky. This is what a red kite would see soaring over the woodland in summer, with all the ground concealed except for rare clearings. A few beech trees have crowns that stand higher than the rest; I infer that one of them is the fine old tree in the middle of our wood, surely a hundred feet tall. Some are laden with the yellow-green cupules that will mature in the autumn. I spot a cherry tree keeping up with its neighbours in the race for the light. To the north, the edge of Lambridge Wood is lined with trees on its margin that have boughs bearing leaves all the way to the ground, like green waterfalls. A huge ash is more delicate; I am reminded of the graceful paintings of Corot, where trees are like gentle breaths.

I have a vision of the whole wood as a tent draped with a canvas of leaves capturing sunlight for photosynthesis. Tree trunks are like immense tent poles supporting the whole structure, and sheltering inside the tent all the animals and plants I have discovered. From this high vantage point the Fair Mile is just one valley in the rolling land-scape, with Henley Park and all the hills beyond, without any jarring buildings until the distance is swallowed up by the most gentle of mists. Henley tucked into its Thameside valley did not look very different in the eighteenth century. Only a conifer plantation in the lower part of Lambridge Wood seems inappropriately dark. I am surprised that there are not more birds up here. I was expecting small species like nuthatches and tits to be hopping everywhere through the high branches. Maybe my presumptions were primed by those tele-vision documentaries displaying abundant life in tropical canopies. Things might be different in the Temperate Zone.

I must not hog the heights. A team of entomologists from the Natural History Museum wants to get up to the canopy to sample the insects that live among the leaves. Shane's machine lifts the visitors up one by one. Their fine nets swish through the canopy, each researcher after his or her own favourites: dipterists in pursuit of flies, coleopter-ists chasing beetles, hymenopterists after tiny insects belonging to the same order as wasps. A dozen swishes, and a skilled collector of tiny insects can gather enough specimens for weeks of work examining the veins on the wings or the hairs on the legs – the stuff of accurate taxonomy. Earlier in the day we had seen the largest British wasp – the hornet (*Vespa crabro*) – buzzing lazily through the trees by the wood-pile. Entomologists are less alarmed by this venomous creature than are ordinary souls; they know it will not attack unless provoked. A second site allows sampling through oak canopy to provide further species, including a spectacularly large weevil (*Curcilio venosus*) that develops in acorns; everyone gathers round to admire its oddly extended mouthparts, almost elephantine. Samples are collected in jars to transport back to the Museum for accurate identification. But one species, *Agrilus angustulus*, immediately sets a beetle man or woman's heart aflutter; for this is a small jewel beetle, a brilliant iridescent green, belonging to a family (Buprestidae) that is much commoner in the canopies of tropical rainforests. Like the yellow bird's nest, it is 'nationally scarce' in the current British classification recording wildlife that is worth protecting.

I would never have been able to find, let alone identify, such interesting items, and my heart rejoices that there are still experts who are able to add them to the cast of characters in the wood. Many more samples disappear with their keepers back to the vaults. The gurus will report back to me later.[10]

Nettle fertiliser

One corner of Grim's Dyke Wood grows stinging nettles. This corner is nearest to the ghost cottage, so it is possible that the ground was disturbed there, since nettles prosper in such places. Nettles provide food for several of our species, including the larva of the charming little Nettle Tap moth, *Anthophila fabriciana*. They are also something of a nuisance, as ours seem to have an exceptionally potent sting. By late June they have grown to full stature. A satisfactory form of vengeance is to be had by turning them into fertiliser. Nettles sequester all kinds of plant nutrients. Strong, thick gloves are needed to pull them up, and then they can be packed into a waste bin with a tight-fitting lid. I scrunched up the leaves and stems as I stuffed them into the bin and pushed down. Now the mush was covered with water (rainwater is good), the lid replaced, and the brew was allowed to rot down. Adding a weight helped. I used a couple of large flints (from the wood, naturally) resting on a piece of wire mesh. Next, I left the vessel in a corner of the garden and forgot about it for a month, perhaps longer. It all fermented in my absence, and the liquor really ponged when it was ready for use. I put a clothes peg on my nose and removed and threw away the soggy stems on to the compost heap. I had to dilute the remaining liquor at least fivefold before using it to water tomatoes, beans and the like. It worked as well as expensive commercial fertiliser.

4

July

Gloom

The rain never stops. The sodden woodland is dark and depressing. Even the birds have stopped singing, except for one bewilderingly optimistic thrush relaying his mantras five times over and over, as if pleading for something. Fungi love damp, but for some reason they too seem discouraged. The only flower still evident is enchanter's nightshade, now showing modest spikes of tiny pink flowers, two petals apiece, and creeping along where the bluebells once were. Who was this enchanter, and why was this little plant his personal night-shade? Where has all the spring exuberance gone? Everything seems portentous. On such a day hardly any light penetrates the canopy. As I look upwards, the leafy roof of the wood might just as well be some unspecified infinity, for all I can see of a junction with the leaden sky. Things merge together above my head. It could be a high mist, or else a trick of the light; I can't tell. Boundaries are smudged out in the wet.

As for the beech trees, rain proves that none are actually vertical. The water always seems preferentially to dribble in rivulets down one side or another of any trunk, darkening it, feeding a million micro-organisms into a brief hegemony. I try to imagine how many amoebae and paramecia rejoice single-celled in the general sogginess. This is a time for mobile microbes to take charge of their slippery fiefdom. The raindrops themselves have allied to form a kind of aggressive bomber air force. All the leaves have long since given up trying to shield those creatures like me who cringe below them. Individual drips now

77

coalesce into giant drops that plummet directly down the back of my neck: huge, unnatural blobs that pass through my inadequate mackintosh. Damp stains spread upwards from the forest floor, and downwards from my collar.

Water is backing up along the paths, so I cut through the big clearing. How much smaller it seems now that the trees are leaning out to their greatest extent. Rain blurs my vision. Arching, vicious bramble stems snag my trousers; I could swear they were deliberately trying to trip me up. I could fall and cut my hand on one of the flints. My own wood suddenly seems to have a malevolent side. I can see how easy it would be for unprepared travellers to lose their way. A sudden, involuntary shudder, and I understand why extensive woods were once thought of as wilderness – unfriendly wastes populated by unreliable sprites. Shakespeare described such a forest in *The Two Gentlemen of Verona*:

> How use doth breed a habit in a man!
> This shadowy desert, unfrequented woods
> I better brook than flourishing peopled towns.

A desert: a place deserted, or a place of desolation, and assuredly not one for quiet contemplation or spiritual refreshment. By Shakespeare's time the wolf was probably extinct in Britain, so the fearsome bogey that inspired so many folk tales in rural Germany was no longer a threat. But dense woods surely continued to seem full of dangerous shadows, some of them real enough, like footpads and desperate fugitives. Even the trees look different this morning: is that the same Elephant Tree? I am sure it was less massive yesterday, and now it is somehow intimidating, even looming in an odd way. That holly scrub is so dark it looks like a hole in the fabric of nature. I lick a bleeding scratch on my wrist inflicted by an old blackberry vine. The salty taste of blood mixes with the blandness of rainwater. This is a strange day, all right. It is a day for delving into the dark past. I think I am alone, but I am not. A lone dog-walker deeply protected by a sensible Barbour jacket and cloth cap is whistling for Rover continuously, and marching along Grim's Dyke, head down, with more urgency than might seem necessary. I cannot see his face. We do not exchange pleasantries.

Grim

The south-western edge of our wood is defined by a bank a few feet high. It is steeper on our side, though hardly dramatically so. Running along beside it, and tracing the length of our boundary, is a shallow depression, a gulley if you wish, about ten paces across. Mature beech trees grow within the gulley, so it is obviously not a new feature. Flints break through to the surface on either side of it, suggesting bedrock brought up by the excavation of a ditch of some kind. It does not look like a particularly important piece of archaeology, and the wood is fairly dotted elsewhere with depressions and banks that are not so different at first glance. However, this gulley extends well beyond our piece of woodland. I have followed it through most of Lambridge Wood where it continues, quite straight, through beech grove and holly jungle, a little more defined in some places than in others, occasionally hard to see, in the direction of the old manor house at Greys Court. It is marked on all the old maps, usually by a double line of hatching.

Grim's Dyke can be alternatively spelled Grime's or Grymes in old sources, and it predates any written record: it is ancient. It also gives its name to our piece of wood, though I suspect that this was mostly a sales pitch to attract the romantically minded (it worked rather well). A few miles away near Nuffield, close to the top of the Chiltern scarp, the same name is applied to a dramatically deep gash that descends straight down the hillside and continues all the way across the Aylesbury Plain beyond to the River Thames at Mongewell, cutting through the present agricultural patchwork as if to demand due acknowledgement of former times. Whoever constructed it, this was a serious piece of ditch-making. Several other comparable structures further north in the Chiltern Hills carry the same old name; in fact, Grim's ditch appears in no fewer than ten English counties.

Nuffield church records show that the dyke there was a familiar land-mark in early medieval times. Between Nuffield and Greys Court other sections of the ditch present a more modest continuation of the same structure, and it is possible to join these pieces together and extend them further into our own wood, making some kind of line extending across country. In spring I have walked along the old foot-paths that follow the dyke above Nuffield, lined with bluebells, as befits an ancient piece of country.

In Norfolk, flint mines known as Grimes Graves were excavated five thousand years ago by Neolithic peoples in search of the perfect natural material with which to manufacture their stone tools. It is that same Grim again. The name dates from early Saxon times. Grim was a title for one of the many guises of the pagan god Woden – Odin of Norse mythology. Grimnir was a shape-shifter, 'the hooded one', a shrouded figure who led souls to the afterlife, a frequenter of dark places. I recall that dog-walker in the wood rushing past in the rain. What was his real business? Had I really seen him at all, or had some holly shrubs briefly created an illusion? Our Saxon forebears knew what the Romans had left behind, but they found many places bearing the scars of far older, mysterious carvings on the landscape: deep ditches cutting across country, odd terraces. To superstitious farmers every nuance of their local countryside must have had hidden mean-ings. What could be more natural than to associate these strange workings with the gods?[1] Grisly Grim might have been an obvious choice, whether for dykes or flint excavations. After all, in later times the Devil himself was tacked on to purely geological creations like the Devil's Tor on Dartmoor or the Devil's Punch Bowl in Surrey. There is an attraction to the diabolical when some inexplicable feature requires a label; it is as if the name alone is sufficient explanation.

Grim's Dyke was dug out and maintained by mere humans, though doubtless beings with their own gods and their own fears. Clearly, the dyke was already old by Saxon times. Richard Bradley has pointed out that not all the 'Grim's dykes' in Oxfordshire are necessarily part of a single system. However, he agrees with other scholars about their age, saying drily: 'We can now accept that there is some body of evidence to favour a context for all these dykes in the later part of the pre-Roman Iron Age.'[2] New evidence has been added recently to confirm his assessment: there is an archaeological feature more than two thousand years old forming one side of our small wood! Who needs spectral apparitions, with or without dogs?

Hard evidence of Grim's Dyke has been lost further down the hill sloping towards Henley. It is very likely that it continued there, because the *Victoria County History* finds reference to the old ditch in documents from the fourteenth century that are housed in the Oxfordshire Records Office. At that time it formed a natural boundary for the old Henley manor of Phyllis (quondam Fillets) Court to the north. Its course finally ran to the River Thames along what it is now New Street (actually a very old street), on the northern side of Henley. It may have been still in evidence much later, for it has been claimed that 'Grimm's Dyke, would at the beginning of the seventeenth century, have been plainly visible in the meadow stretching behind the brew-house in North Street, facing the entrance to New Street.'[3] All this land is now built over. If the evidence is totted up we finish up with a ditch that runs from Henley to Mongewell for more than ten miles over the scarp of the Chiltern Hills, connecting across the great southerly loop in the course of the River Thames. Some pieces of the ditch are missing, but these are often in places where the land has probably long been under the plough. Farming slowly rubs out time's messages.

What was Grim's Dyke for? I return to the wood for some detective work. It is really not much of a trench here, perhaps twenty-five feet or so across, with a higher flinty bank on its western side. This must have been thrown up when the feature was dug out. The depression will have filled in over two millennia, but the stony ground seems to have retained its form. It is too small and the wrong shape for any kind of defensive dyke – such a structure normally would have a steep wall facing any potential enemies. The archaeologist Jill Eyers came with her team of volunteers to dig a trench across the dyke to root out any evidence that might remain there. It did not yield much to their mattocks and spades – no coin or pottery fragment that would have made the excavation an event. Our section of Grim's Dyke is certainly no Maginot Line, and is much less of a barrier than the striking gulley that runs down the Chiltern scarp. I conclude, as have many others, that the dyke here must have been some sort of marker, perhaps a territorial boundary. Earlier historians would have made it the border between two tribal groups recognised by Roman chroniclers: probably Catuvellauni to the east, Atrebates to the west. Modern writers are more cautious, and recent interpretations have even mentioned the ditch as fencing in an extensive cleared area for raising livestock. All authorities agree that the dyke (or dykes) is connected with the Iron Age hill forts of the last millennium BC.

Hill forts are distributed quite regularly along the high edge of the Chiltern scarp. On the ground they are marked by prominent concentric ramparts surrounding a central enclosure. There is usually one defensible entrance to the enclosure, within which the postholes for collections of round huts can usually be uncovered using trowels and patience. When they were built they were situated on extensive areas of cleared ground, with far views to the north, but nowadays the forested Chiltern hilltops have often reclaimed the old forts and sequester them inside deep woods. More of them are being discovered as this is written. Away from the scarp slope there is a similar structure guarding an ancient crossing over the Thames at Medmenham, about three miles downriver from Henley.

The hill forts are in their turn related to trade. One of the most important routes in ancient Britain runs along the base of the Chiltern Hills where it joins the plain to the north: the Upper Icknield Way. I cannot better H.J. Massingham's description of 'the old road on its journey from Norfolk to Devon. It has crept along the ankles of the shaggy range, below the trees but above the springs, just where the upper Greensand borders the chalk.' Skins, wood, iron ore, exotica from the Roman Empire all traded along a route dictated by deep geological structure. Grim's Dyke crosses it on its way to the Thames at Mongewell and the river. This whole area must have been abuzz with cowherds, farmers, shepherds, traders, soldiers and shamans. The population at that time was increasing. The hill forts commanded a view of all this activity, and doubtless extracted or traded their share of it. Grim's Dyke is not simply an extension of the fortifications. Instead, it traces an independent course, as might be expected of a boundary. It may have demarcated the sphere of influence of one or more chieftains, masters of the forts; they must have been powerful figures to coordinate the ranks of ditch-diggers required for the job. Perhaps this was the first time the iron fist was brandished.

By way of which history we arrive back at the wood. If Grim's Dyke was once a boundary, I have to conclude that there *was* no wood! To bury a border of whatever kind deep inside a beech forest makes no sense. It should be out in the open. When the hill forts and their surrounding areas were cleared of trees there was a growth in farming of all kinds as the number of hungry mouths increased. Clearing was general. In the later Iron Age the climate was warming up after a long cold period, and this too supported greater agricultural production. The higher ground may well have been easier to clear than thickly forested and poorly drained plains, and no doubt sharp iron tools were helpful in that task. From our part of Grim's Dyke Jill Eyers recovered pollen of arable crop plants. So the centuries before the Roman invasion provide a baseline for our wood – a time when it wasn't there. Grim's Dyke Wood was more likely Grim's Dyke Down. I have a vision of men at work, digging the dyke, dressed in simple leather jerkins and trousers, the crack of iron on flint resounding in the air. Maybe some women dressed in simple flaxen shifts bring in a scattering of sheep from the surrounding fields. It is a scene full of light, not darkness. The language they speak is half-familiar from my days working in the Welsh mountains – some kind of guttural Celtic tongue, neither Breton nor Gaelic, but an ancestral version. I even feel a brief tug of affinity. When the DNA of my Y chromosome was analysed a few years ago in a laboratory in Oxford I was declared to belong to a Celtic population (on the male line).[4] I was no Viking, Saxon or Roman. Those ditch-diggers were my people.

These speculations are garnished with purest gold: a glittering hoard buried near Lambridge Wood. In 2003–04 a sleuth with a metal detector discovered a treasure of thirty-two gold coins in a field just to the west of Grim's Dyke which was the best trove of its kind ever found in Oxfordshire. They were concealed inside a hollow flint – placed there for safety, or even as a gift to the gods. The coins were minted in the 50s BC, near the end of the late Iron Age. Both flint and coins are on display in the River and Rowing Museum in Henley. Gold never tarnishes, so these precious discs are as brilliant as the day they were made. Each coin is about the size of a US quarter and has a beautiful, schematic 'triple tailed' horse stamped upon it, standing over a perfectly suggested chariot wheel. The reverse side is unmarked. Coins of this type are attributed to the tribe of the Atrebates, and indeed the find was made on 'their side' of the dyke. They may have been minted at Silchester (*Calleva Atrebatum* of the Romans), an important Iron Age city only fifteen miles or so to the south of the wood. Numismatists are able to match these 'Celtic' coins all the way from the Danube to northern France, until they were created in ancient Britain itself in 120 BC. The design was copied and recopied from a pattern that originated in Macedonia before 300 BC, depicting the two-horse chariot in which Philip the Second of Macedon triumphed in the Olympic Games of 352 BC. Here on the edge of Lambridge Wood are golden tokens that marry the Chiltern Hills with one of the cradles of Western civilisation.

I admit that I have been shaking appropriately shaped hollow flints ever since I heard about the hoard. I have failed to discover treasure (so far), but I have found some remarkably globular flints. Curious stones like these have attracted attention for a long time. In the *Natural History of Oxfordshire* (1677) Robert Plot writes: 'Here also must be referred a *Round Stone* ... containing within it a white sort of Earth and therefore called *Geodes* or the *pregnant Stone* ... the outer Crust of these is sometimes on an indurated *Chalk* ... and when found thus, by the inhabitants of the Chiltern (where they are most plentiful) they are called *chalk Eggs*.' Only one searcher has been lucky enough to discover a *golden* egg. The spherical stones have a down-to-earth-enough explanation. They are flints that formed around ball-shaped fossil sponges (*Porosphaera globularis*), and were originally derived from the chalk. Like other flints, they survived the erosion of the soft limestone that once enclosed them and nowadays may be brought up by the plough, or occasionally scattered on the surface, including the ones I pounced on in our wood. The white sort of earth in the 'pregnant Stone' expounded by Plot is often all that remains of the fossil sponge inside the flint. With further weathering the flints hollow out completely, and then they make a suitable receptacle for a cache of gold coins. One particular Iron Age individual picked up a large example of such a near-immortal purse, and used it to tuck away treasure. To appease my covetous fantasies I shall have to be content with adding just three small 'chalk Eggs' to the collection.

There is no evidence in our wood for still earlier times – the days of the wildwood and after, more than eight thousand years ago. I will never know when the first footsteps crunched over the surface of our clay-with-flints. I have extracted fresh flints from the chalk beyond the Fair Mile, and they have wonderfully dark and uniform interiors; just the kind of flint that was used to manufacture sharp-edged stone tools. But I have never found any ancient implements of Mesolithic or Neolithic age in all my traverses across Lambridge Wood.[5] In any case, there is enough known from the surrounding Chiltern Hills to show that human beings used the resources of our area long before the Iron Age. Mesolithic tools have been found as close as Stonor just up the Assendon Valley. A site at Chesham twenty miles to the north-east of us has yielded many tools of the same general age along with bones of red deer, boar and the extinct wild cow (ancestor of our domestic cattle) known as the aurochs (*Bos primigenius*). This is enough to conjure a vision of skin-clad hunters questing through thick forests (and our wood) stocked with edible but wary beasts. In turn, these people must have dreaded the attention of wolves and bears. Theirs was an opportunistic life, often on the move, settling around a rich seam of flint for a period before moving on once more. A dropped arrow, a damaged scraper for dressing skins – these were the only traces they left behind, their momentary lapses preserved for eternity thanks to the indestructibility of flint. Iron swords a fraction of their antiquity have become unrecognisable clots of rust.

Permanent settlement began with Neolithic farmsteads (3500–4000

Aurochs from Les Eyzies, France.

BC). Caring for domesticated animals and the cultivation of early grains such as emmer wheat demanded clearance of patches of the wildwood. New livelihoods stimulated new skills. Pottery utensils for storage of foodstuffs for winter use and everyday cookery began to bequeath the scraps and sherds that provide the archaeologists' bread and butter. Important persons were interred with grave goods in burial mounds – the barrows that are dotted over the English countryside. Centuries of ploughing have obliterated many, but in dry summers their outlines still emerge like repressed memories, often in the midst of cornfields. Areas underlain by chalk were favoured for clearance, but there is no Stonehenge or Avebury in the Chiltern Hills. Such large ceremonial centres lay on what are now the open chalk plains of Wiltshire to the south and west rather than the comparatively rugged Chiltern country. The only important sites yet discovered in our range lie in the northern section of the hills around Dunstable.

The Upper Icknield Way was, however, already an important route, particularly as trade developed, so there must have been people passing through our patch of country. A greenstone adze has been found nearby which was traded all the way from Scotland. There is evidence of Neolithic settlement near Wallingford on the River Thames, just to the north of the point where Grim's Dyke reaches the water. I am certain that there would have been forays into the hills to collect those beautiful, black-centred flints, so good for the manufacture of sharp blades. 'Cutting-edge technology' was already an apt description this early in human history. I imagine hunting parties, armed with their new flint tools, setting out for the hills in search of game: the thrill of a deer brought down, the welcome the hunters received on their return to camp.

When flint was succeeded by bronze (about 2500 BC), clearance of woodland probably proceeded further. John Evans has proved in the northern part of the Chilterns that the small species of snails recovered from excavations changed through time from specialists favouring life in woodland to those that thrived better in open grassland.[6] Their shells remain behind to reveal such details. For these areas this change marked the first time cattle and sheep grazed the same hillsides that would support them for thousands of years to come.

If the general history of landscape before the Iron Age is known, our piece of woodland has no precise place in that narrative. It is as well to recall the old aphorism 'Absence of evidence is not evidence of absence.' The lack of Neolithic flints from Lambridge Wood is no proof that hunters avoided it. A beautiful Bronze Age sword was recovered from the river near Henley-on-Thames and is now on display in the River and Rowing Museum, but if ancient braves ever wandered through our wood they left no trace. We do know that there were people in our area from Mesolithic times onward, and that later a major route lay near the foot of the scarp not far to the north; we also know that an important clearance was likely to have happened before the Roman invasion. It is probable that our woods were never left entirely to themselves. I cannot say exactly when the wildwood was erased, never to return. We do not even know exactly what the wildwood looked like. Oliver Rackham estimates that about 80 per cent of the British landscape was mostly tree-covered at 800 BC, which generally favours the idea of later clearance. The absence of any

Bronze Age barrows near our wood would also indicate a later, Iron Age date – but again there is no proof. What I can say with certainty is that the subsequent regeneration and persistence of our wood helped to preserve the evidence of the Grim, Grimm, Grime's ditch – choose whichever strange devil you like – by protecting it from erasure by plough and farrow; the churn and churn again of agriculture elsewhere sponsored slow amnesia in the landscape. Old woods are places with a much longer memory.

Yew

And even within the wood, some trees have longer memories than others. The most enduring tree of all is the yew (*Taxus baccata*). The oldest yews in the world are found in Turkey, and are estimated as over three thousand years old,[7] which on a scale of our wood's history takes us back well beyond the original excavation of Grim's Dyke. The age of the oldest British yew is disputed, but one vast tree at St Cynog's church in Powys, Wales, has been claimed to rival the Turkish trees in antiquity. So the yew is a living bond with our era of mystery. Just above the Icknield Way, on Watlington Hill, veteran yews line one of the old tracks; their ribbed trunks are fantastically bent and twisted, sere branches droop to the ground and even take root, and dryads peep out from every inscrutable cranny.

Our own two small trees in Grim's Dyke Wood are just at the start of their unhurried saga. Their crowns are well-formed cones of branches, and each branch carries splays of leaves divided into not-quite paired, flattened, tough, deep-green needles. From afar the little trees – whose trunks I can almost encompass with my hands – look implausibly dark, almost black, especially when the beech leaves are at their brightest to supply a shimmering backdrop. Yews don't mind growing in the shade; they revel in it. They seem almost a part of it. They have a metabolic trick that helps them thrive as understory. During the winter months, when light floods the wood and everything else in the plant world has battened down to see out the season, yew leaves continue to photosynthesise. They build their strength while others sleep. The trees grow fastest while they are at our young, *moderato* stage, later slowing to *maestoso*, finally settling into *molto lento*, with growth rings much less than one millimetre wide. I guess that my trees established themselves from seed after the big felling eighty years ago. They still have a long way to go, but when the oldest beech tree falls the yews will not even have reached middle age.

Yew is an extremely poisonous conifer. Cattle occasionally nibble at the fresher green foliage that appears late in spring from the tips of the branches, and then they die. This raises the question of how the seeds get distributed, since they are equally poisonous. The only part of the tree that is *not* toxic is the 'fruit', which is a bright-pink, fleshy cup the size of an orange pip, very conspicuous in autumn. It is a strangely modified cone – strictly termed an aril – that surrounds a single seed; the inconspicuous male cones are carried on separate trees. The arils on their own are reputed to taste delicious. Proving this is one experiment I regard as beyond my research remit. Thrushes eat the arils and the seeds together, but the latter pass through their digestive system unchanged, and ready to germinate in a new site. Badgers are supposed to manage the same trick. If we were to scoff whole just a handful of these tempting pink 'candies' we might well not survive. But like many poisonous plants, but very few hollow flints, the yew hides secret treasures: this tree is the original source for a group of drugs (taxanes) that are very useful in slowing or stopping the growth of a variety of cancers. These drugs are one of the most prescribed of all chemotherapies. Huge quantities of yew bark were formerly used in their manufacture, but fortunately, artificial synthesis of relatives of the active compounds has reduced excessive demands on the wild yew population in the last twenty years.

There is nothing new about threats to the survival of this remarkable tree. Longbows were the most effective weapons of the fourteenth to sixteenth centuries. They had to be made of different strips of yew: the heartwood closer to the archer compresses as the bow is drawn; conversely, the sapwood forming a layer on the back of the bow is elastic and lengthens during use. Only yew wood has these extraordinary abilities, which, properly exploited, could propel arrows

Fifteenth-century misericord, St Mary's church, Beverley, Yorkshire.

through chain mail. Welsh archers were said to be able to impale fully-armed enemies by piercing their armour, leg, saddle, and underlying horseflesh in a single shot; the late medieval equivalent of Clint Eastwood's achievements with a .44 Magnum (except probably true). The bowyer was an exquisitely skilled craftsman who instinctively understood elasticity, aerodynamics and the limitations of human musculature in making instruments of warfare; the arrowsmith shaping his goose-feather flights was scarcely less adroit.

In northern France on 25 October 1415, twenty to thirty thousand Frenchmen, including mounted knights dressed in masterpieces of the armourer's trade, were defeated by five thousand archers skilled in the deployment of yew longbows, backed up by a mere nine hundred men-at-arms. The Battle of Agincourt was the major English victory in the Hundred Years War. It is no wonder that English demand for yew was relentless, nor that it made for big business. The local supply was soon insufficient. It is as well for the survival of old English trees that many venerable yews grew in churchyards, and were inviolable. In 1473, Edward IV decreed compulsory yew imports. For well over a century yew wood was sucked in to London from all over Europe. During the years 1512–1592 the Austrian company of Chrisoph Fürer

& Leonard Stockhammer exported 1.6 million yew staves, and this was just one company among many. The combination of the slowest-growing tree with the fastest-rising demand was obviously unsustainable. The forests of Bavaria and Austria were stripped beyond any capacity for natural regeneration. Even Carpathian yews were obliterated to supply the greedy longbows. It could not continue. On 26 October 1595 Queen Elizabeth I decreed that henceforth the army should replace its longbows with guns, even though the bow was still much the more effective weapon. The few remaining wild yews could resume their slow journey towards immortality. Our own small yews still have a thousand years in hand.

Deer and dogs

I hear the strangest cries. They sound like panicky barking from a dog separated from its master or mistress. But the noises are not as insistent as I would expect from a lost dog. Each bark is separated by a few minims of silence. There is something of the crow's hoarseness about it, but I know it is no bird. The high mewing of a red kite is the only bird I can hear today. Then I spot a warm-brown muntjac deer (*Muntiacus reevesi*) picking its way delicately through the big bramble patch. Despite its assertive bark, it is a shy animal that is able to vanish in a trice into the dappled distance, so I try to remain invisible. Through my binoculars I observe a pair of tiny antlers, pertly erect, flanked by big sensitive ears. I can just make out a pair of sharp 'tusks' projecting down from the deer's upper jaw. The whole appearance of the animal is somehow defensive, which may be because of its rounded haunches, which seem to make it cringe. It does have something dog-like about it, after all – an unhappy dog. The little muntjac delicately browses a hazel leaf, plucking it fastidiously, chewing discreetly. Then it picks up my presence by some sort of nervous telepathy and, momentarily flashing a white patch beneath its tail, it is off. I have no idea how many muntjacs there might be in Lambridge Wood, but I know there must be a minimum of two, because later there are two barks coming from different directions.

No wild mammal eats yew, but almost every other plant in the wood is a potential snack for deer. The solitary muntjac is the species I see most regularly. Twice I have surprised a group of roe deer moving in jumpy concert through the wood. They are quite the prettiest deer; when they turn to appraise the threat from a human they show a great black smudge of a nose and genuine doe eyes. The male's antlers bear just a few elegant spikes. There is nothing apologetic about their haunches. For a moment I might be that Neolithic hunter after wild game: roe deer were denizens of the wildwood. Bears, wolves and aurochs may have been exterminated or banished, but roe deer still move quietly through copse and woodland, probably safer now than they were three millennia ago. The look they give me is an ancient look, one of wariness stored in the genes and worked out in nature. I do not tell them that I have the shooting rights in my wood, nor do I add that I am probably the worst shot in the northern hemisphere. I am also clean out of stone tools.

I know for certain that fallow deer, too, have been through the wood, for I have found a shed antler. It is a grander affair than the roe deer's, with a flared crown and a prominent tine near the base. It is too large to be added to the collection. I have no evidence of the rut from my patch, but I have noticed scuffed patches on dead branches that mark where deer have rubbed off the velvet from their antlers. The Plantagenet Edward, Second Duke of York, hero of Agincourt and author of the hunting treatise *The Master of Game* (1406–13), described this ritual thus: 'about Mary Magdalene day [22 July] they fray their horns against the trees, and have rubbed away that skin from their horns and then wax they hard and strong'. Here was a man equally versed in the histories of harts and yews.

The three deer species in our wood have different claims upon the landscape. The roe deer is the ancient inhabitant. Fallow deer were introduced by the Normans for sport and for meat; some say the Romans did this long before. The muntjac is a Johnny-come-lately from China that escaped from Woburn Park and elsewhere early in the twentieth century – a story rather like that of the grey squirrel. A colleague tells me that its odd teeth indicate that it is close to fossil species from the Miocene period, so in one sense it is the *oldest* inhabitant. All three species do very well browsing in our wood. They love hazel foliage and other fresh leaves. The muntjac deer may be the main culprit in nipping off the flowering spikes of rare orchids elsewhere in Lambridge Wood. If the main interest in woodland were preserving flowering plants it might be better to exclude deer altogether, but I hate the thought of fencing around our little plot of trees. All naturalists and ecologists (not to say farmers) agree that there are far too many deer roaming freely through field and forest, but there is no consensus about what should be done about it. Most people draw the line at reintroducing wolves.

The fortunes of deer in England have changed. After the Norman Conquest deer parks and royal forests were the apogee of regal and aristocratic distinction. Wonderful medieval tapestries in the Musée de Cluny in Paris show noble stags and hinds treading on carpets of flowers arrayed like scattered stars. The deer hunt was more than sport; it was the embodiment of kingship. William Rufus introduced dreadful punishments for any infringement of Forest Law, and poaching venison – or indeed anything – became a risky venture. Stonor House, three miles north of the wood, is still surrounded by a deer park. Eight hundred years of occupation by the same family encourages a certain respect for the past, though the large herd is no longer hunted with bow and arrow; it is culled when necessary using suitably modern means. A deer park was an important part of Greys Court in its medieval heyday. For centuries, venison was among the most highly regarded of meats.

Recent research suggests that the few surviving ancient parklands may approximate to a kind of open forest – more savannah than dense wildwood – that once spread widely across north-west Europe.[8] Deer parks continued to be a part of the necessary trappings of large estates, to demonstrate success and status to envious neighbours. Private grounds became less coarse pasture with scattered trees, dedicated to a thoroughly practical function, and more a display for vistas designed for aesthetic qualities. But even the picturesque and deliberate landscapes of the eighteenth and nineteenth centuries liked to feature elegant deer. Henley Park, a few hundred yards from the wood on the other side of the Fair Mile, was then not so different from Stonor as it is today. Cattle could be an attractive grace note on the planned landscape, but handsome bucks still had the edge. Venison continued to be a staple with status for the table, although only the leg and part of the saddle was served to royalty. The Duke of Wellington's favourite dish in 1816 was neck of venison.[9] Allowing fine bucks to go to waste would have been inconceivable. Nowadays, battered carcasses of deer knocked over by speeding vehicles are a common enough sight in the Chiltern Hills. Nobody stops to pick up the bodies. Magpies peck at them, fitfully. What might once have fed a family for weeks now rots unheeded by the side of the road. There are too many deer, and they are held in too little regard.

Deer cause damage to woodland by topping off seedling trees before they get established. I have met wood-owners who complain that they have no natural regeneration whatever as a result of such depredation. I was puzzled why Grim's Dyke Wood seemed to have escaped the worst. There are dozens of seedling ash trees all vying for a place in the canopy, and plenty of young beeches, and even small cherry trees. I am more worried about squirrel damage than I am about the destruction caused by deer. On one visit in July I realised why this should be. Four roe deer ran past me in full flight from the section of Lambridge Wood further down the hill. A few minutes later a professional dog-walker emerged along one of our footpaths with a straggling bunch of dogs on leads, and an aged retriever plodding behind. Several dogs cocked their legs against beech trees as they passed. As far as the deer were concerned these were wolves, in a pack, and their Neolithic flight reaction was immediate.

Grim's Dyke Wood has three public footpaths running more or less around its perimeter, and it is easily accessible from Henley-on-Thames. It is part of a regular beat for dog-walkers; the paths must be redolent with dog scent. Deer coming this way will always be twitchy and on guard. This may explain how our young and regenerating trees have survived. I wish that squirrels were equally vulnerable, but they scamper up trees with insouciant ease to safety when enthusiastic puppies that know no better attempt to catch them. In general, though, dogs are my friends.

Gratitude was not a general emotion early in my wood-owning days. I may be a dog-lover, but I do not feel obliged to love their owners. An Alsatian comes at me, barking dangerously and showing its teeth in a way that does not require an ethologist to decode. 'He doesn't like people wearing hats,' the middle-aged woman who owns him says accusingly. Or I am way off the footpath searching for fungi underneath the holly trees when two mongrels attack me in a way that would cause a roe deer to die on the spot. 'They don't like people carrying baskets,' the gum-booted couple explain haughtily, whilst retrieving their animals with an air of reluctance. 'Particularly under bushes.'

Into the sunshine

It is good to have breezy, sunny weather after all the rain. Now the wood displays brightly illuminated patches on the more open ground, while under the thickest beech canopy the flinty soil remains dark. There is a kind of dancing chiaroscuro all around as the wind tosses the high beech foliage. Everywhere is a whispering susurration. New leaves on the hazel are held out like hands gratefully to receive the light around the clearing. I am glad to see that the squirrel damage so conspicuous a month ago has faded to grey. Seen from afar, cherry trunks rise blackest and straightest in all the wood, while their polished bark almost glitters in the sun, but the same paintbrush dapples the trunks of the beeches in gold flashes. The lower beech branches wave slowly in the breeze, back and forth like seaweed fronds under water.

Tufted hair grass (*Deschampsia caespitosa*) is the last of the woodland grasses to flower. It is all elegance, carrying several three-foot-high sprays that glisten silver in the intense light like feathery fireworks. False brome (*Brachypodium sylvaticum*) is a leafier grass with a shorter, drooping inflorescence, as if the effort of flowering is just too much. The sun has brought out white bramble flowers held above the big patch, but some early fruit has already set, all dense and green surrounded by a tonsure of brown and faded stamens. I shall never know which of the four hundred bramble microspecies is our very own; no flower book can tell me. But whatever its scientific identity, the white blossoms that seem rather routine to us must have a very particular attraction for insects. Here are hoverflies of at least four kinds, dashing and pausing, some imitating wasps, others pretending to be bees. With two rather than four wings they are only flies in fine dressing. A comma butterfly is no more than an orange flash until it briefly settles to reveal the etched and scalloped edges of its wings. A red admiral flashes its scarlet bars of authority as it unrolls its proboscis carefully into the white florets. It seems to be in no hurry at all. Several flitting, flirting ringlets are all dark-brown discretion until they close their wings and reveal several small white 'eyes' with bright dots at the centre on the undersides. A green-veined white butterfly looks so much more decorous than the white butterflies whose caterpillars destroy my cabbages. They are all as smart as can be. Then just a single silver-washed fritillary, much larger than the comma, almost glides in and settles – the prize of the show. Like a mannequin it twirls around a flower head to show off its orange livery, all spotty and dashed dark-brown. I will it to stay, but it is off to wow another naturalist.

The breeze has brought down clumps of cherries from the crowns of the trees. They hang together, a few cheery red balls like the chubby cheeks of cherubs, the biggest not much more than half the size of the cultivated variety. I taste one carefully. There is a gentle, acid sweetness to the flesh that is much more subtle than the lip-smacking lusciousness of the commercial fruit that descended from it. A few untidy clots of pale cherry stones are scattered around on the ground. They must be the excreta of some cherry-guzzling bird that has been feasting on the brief bonanza high above my head. But which bird? Surely only a big, black crow would have the craw to crop a crowd of cherries.

In spite of the relatively recent rains this hot, dry weather has turned the beech litter into a crackling and inhospitable waste. The mosses that were so lately spongy, verdant cushions have now shrunk in on themselves to conserve water. Micro-organisms that flourished in the wet will no doubt have transformed into tiny resting cysts to see out the drought. In such weather small mammals must hunker down in their deepest burrows or under damp logs, only emerging at dusk. I can hear little birdsong now – nothing to detract from the rustling of the leaves, and somewhere in the background the mournful groan of a jet aeroplane finding its way to Heathrow. As I look for fungi in the conifer woodpile I find instead a fat brown toad sheltering under a rotting plank. He regards me with a kind of stoic indifference. I replace his shelter, mumbling an apology.

Wild cherry jam

It is not easy to gather a quantity of wild cherries if the crows are determined enough to get the lot, but after this windy day many little clumps have found their way to the ground, inviting me to make something from them. I know that they do not all have to be fully ripe and red – those that are an attractive apricot colour are perfectly fine for cooking. Cherry stones (pits) contain a small measure of a cyanic compound, much like bitter almonds, but I cannot find any evidence that there is enough to cause any harm. The stones also provide pectin to help the jam set, so if all the stones are removed before cooking, jam sugar containing pectin must be used instead. This is probably the more reliable method. The stones have to be removed from the flesh, tapped with a hammer and placed in a tied muslin bag if they are to be employed as a source of pectin. Some recipes recommend cracking the stones, extracting the kernels and cooking those with the flesh, but this is very fiddly. The quantities of ingredients depend on how many cherries can be gathered, but the sugar should be about 75 per cent of the weight of the cherries. The stoneless cherries go into a pan (with the stones in a bag if they are the pectin source), and a good squeeze of lemon juice is added. I then cook them until the fruit is very soft. At this point the sugar has to be added and the whole brought to what cookery books always describe as 'a rolling boil', which sounds like a medical condition. The setting point is reached in ten minutes or so. If the mixture is reluctant to set it can still be used as a delicious dressing for ice cream.

5

August

Thunderstorm and after

At first the wind whips up the leaves into a squally frenzy, as if some inundation were rushing in to swamp the wood, accompanied by a sound like crashing waves. I briefly close my eyes and I am right by an ocean, and each pulse of wind is another breaker that passes across the canopy. The pull of wind on leaves exactly mimics the roll of breakers over pebbles, that hubbub of motion, as one pulse of energy melds with its successor. The mew of a red kite in the gusts is so like the cry of a seagull. Grumbles of thunder somewhere to the north are followed by flashes of lightning and a mighty crash, apparently coming from the sky immediately above me. The storm is so close that I am sent scurrying back to my car as heavy rain begins to fall. It is safer to sit this one out. Fortunately, I do not have long to wait.

The thunderstorm passes as quickly as it arrived, and then the sun plays through the trees with sudden and exceptional clarity. As the leaves are teased by the breeze their shadows dance with something of the constrained unpredictability of flickering flames. The effect might be described as kaleidoscopic, except that there are only two colours: green and gold set against the dark wall of holly. The high tree crowns rather shimmer in the intense light. A portrayal of the wood as 'magical' does not seem overblown for once, as if Peaseblossom, Cobweb and their fairy friends might suspend the laws of physics here just for a few minutes. A girl on a pony passes along the path, the pair sashaying together like a strange dark ghost through the shade, and hardly visible. Implausibly large glistening raindrops still hang off twigs, like pearls. Newly flooded puddles provide dimpled and approximate reflections of it all.

Rain brings out large slugs from their moist hiding places in the shadier part of the wood. The biggest of them all is a great black slug, *Arion ater*, which is longer and chunkier than my middle finger. A deep-orange variety of the same species somehow looks more unwholesome as its mucus coating glistens in a sunbeam with lubricious succulence. I turn one slug over to watch the involuntary contraction of the foot on which it glides, a contraction that momentarily almost halves its length. Its foul taste is a guarantee that no predator will regard it as a treat, for all that it is a mass of muscular protein. After it is restored right-way up, within a few seconds its two pairs of tentacles re-inflate, the longer, upper optical tentacles orientating the confused mollusc in a trice, while the lower sensory tentacles will soon 'smell' out a mushroom or decaying plant stem for luncheon. Off it glides, purposefully. Another large, but more slender, greyish slug is distinctly black-spotted towards its front, more black-striped behind, like a pinstripe gone wrong. The leopard slug (*Limax maximus*) is no predator, but unlike its eponym *can* change its spots – or at the least is very variable in colour. I have found its spherical eggs clustered under rotting logs, resembling a pallid brand of caviar. Once I saw its crepuscular mating dance, during which a pair of slugs hang from a twig by a thread of mucus, creating a *pas de deux* of unsurpassed sliminess, a sight not everyone would find erotic.

These vagrant molluscs remind me of their relatives, snails – or rather, the lack of them. Not one *escargot* worth marinating in butter has been found in Grim's Dyke Wood. Even the common and variable grove snail (*Cepaea nemoralis*), which is often as brightly striped as a Henley Royal Regatta blazer, makes only a desultory showing at the edge of the wood. I have been on snail hunts and found only minute species. Under rotting logs the sole snail commonly found is a tiny form called the round disc (*Discus rotundatus*), whose name, whether in English or Latin, describes its appearance very well. Seen under a hand lens it is very pretty, ribbed with growth lines, and marked with regular brownish stripes. The glass snail (*Oxychilus cellarius*) is hardly larger, and so thin-shelled it is practically transparent. A hairy snail (*Trochilus hirsutus*) is decorated with a fine fuzz. The total species list is only eight.

In my garden, not so far away from the wood, stout common garden snails queue up to graze my vegetables, and no *Hosta* would stand a chance of growing to maturity unchewed. I was puzzled by this apparently inexplicable contrast, until I worked out the geology. Big snails need calcium carbonate (limestone, chalk) to make their shells. The thicker the shell, the more lime is required. The veneer of clay-with-flints in the wood has all the lime leached out of it. Only the thinnest-shelled, tiniest snails can get the mortar they need to build a home. Soft, unpalatable slugs have no such prohibition, as they do not need to build a shell.

To test my hypothesis, I visited the Hambleden Valley just down-river from Henley in Buckinghamshire, with its lovely chalky land-scape. A footpath at Pheasant's Hill leads up into airy woodland. On the pathside, chalk-loving plants like old man's beard (*Clematis vitalba*), dog's mercury (*Mercurialis perennis*) and the charmingly named ploughman's spikenard (*Inula conyzae*) soon told me I was on a differ-ent soil from that in our wood, even though both beech and ash trees were as abundant. I confess to a moment of wood envy – I really liked this patch of sloping ground. There were also empty snail shells, big ones, to left and right, all belonging to species absent from our wood. They obviously had no trouble finding what they needed to secrete their homes. There was even a thick-shelled, ornamented and turreted form, the round-mouthed snail (*Pomatias elegans*), that had enough calcium carbonate left over to make a protective door to cover its aper-ture. The mystery of the rarity of shells in our wood is really no enigma: lime is just one of the hidden controls on what lives where. Grim's Dyke Wood simply does not have enough of it to make big snails.

Familiarity does not breed contempt; rather, it breeds discrimination. On my perambulations through the wood, now all freshened after the thunder, I love to plod around on a ritual beat with my eyes swivelling left to right, looking for anything that has changed since the last time. This slow walk is never exactly the same; there is always something new to recognise. Near the south-eastern corner of the wood, one of our largest beech trees is surrounded by a natural mossy garden on a gentle bank. Growing out of the moss today is a stinkhorn (*Phallus impudicus*). It was not there a week ago. As with the round disc snail, its Latin name provides a precise description of the organism: it arises from a white sac, which could be appropriately described as scrotal; above this rears a pallid erect member of average (occasionally generous) penile dimensions, the whole capped with a *glans*, but one decorated with greenish slime. *Phallus* is only the first part of the name; as for the second, *impudicus* means 'stinking' (in Latin), so what I have before me is a 'stinking phallus'. And it really does smell: of something ripely rotten, like bad meat or dumped offal perhaps. The stinkhorn is one of the most curious productions of nature; and it is a fungus.

Down on my hands and knees I soon find a second individual at an earlier stage of growth, snuggling in the moss. It looks exactly like a white, round egg, about the size that might be laid by a particularly ambitious hen. It is quite firm to the touch, though the white skin feels a little flexible, like pigskin. When some of the moss is removed I reveal that the egg is attached to a thick, somewhat elastic white thread that runs horizontally near the surface of the soil. This is the business end of the fungus, a mycelial thread that connects the fruit body with the network of fine filaments that feed off decaying wood and leaves in the surrounding woodland. So the stinkhorn arises (I am tempted to say gets an erection) from the egg. Its stem quickly extends and carries the pyramidal 'cap' aloft, smothered in green gunk.

Then I notice two or three fat flies dining enthusiastically on this substance, sucking at it with their mouthparts, like cartoon gluttons slurping soup. The pong is explained. These are carrion-loving flies, connoisseurs of rot and decay. This is their meat and drink. They smell a rotting shrew or mouse. The fungus has fooled them into dining on its microscopic spores, which they will spread through the wood in their droppings.[1] Once smelled, the fetid stinkhorn's aroma is not easily forgotten. I have several times tracked down what Sacheverell Sitwell called the 'horned god of the coven' from catching a whiff of it by the wayside. Curiously, at the egg stage there is no smell. The baby fruit body crouches like a nut inside the egg, surrounded by faintly-coloured jelly. The crunchy bud is edible raw. It creates quite an impression on a fungus foray[2] if, after pointing out the mature fungus in all its glory, the foray leader scoops up an egg, peels out the stinkhorn embryo and pops it into his mouth.

Romans and after

I have explained why I think the wood did not exist in the late Iron Age. I would like to discover when the distant progenitor of the wood that now covers Grim's Dyke first began to transform the landscape. I was brought up with a naïve tradition that the Roman Conquest of AD 43 brought civilisation to Britain, and I admit that my vision of the average Roman was probably heavily influenced by having to study Caesar's *Gallic Wars*, Book 3, for my school examinations, and by several toga-rich Hollywood confections, probably featuring Charlton Heston shouting 'Hail!' to other Romans almost as handsome. What is now known of the earlier tribes who peopled our landscape proves that they had many attributes associated with advanced societies, including complex trade links that already rooted deeply into continental Europe. Most historians accept that the conquering Romans had the advantage of being militarily superior to everyone else, and highly organised; that there were battles that quelled important Iron Age redoubts like Maiden Castle in Dorset is proved unequivocally by the discovery of British skeletons still carrying direct evidence of the hardware that killed them. Our Celtic tribes in the Chiltern Hills may well have delayed their oppressors for some time before they were subdued; it is unlikely that the old chiefs would have compromised a clear prospect over the hills to spy out the enemy – no new trees to conceal the view.

A less violent transition is recorded in other sites where 'Romanisation' may have happened progressively over many years.

The Romano-British as a whole were a people transformed as much as conquered. At the time of the construction of Hadrian's Wall in AD 122 this process had probably advanced as far as it could. By then, a new Roman road had been constructed running straight on lower ground parallel to the ancient Upper Icknield Way at the foot of our local Chiltern scarp, with villas all along the route. By the second century, villas were also occupied at locations scattered variously over the Chiltern Hills, and at special sites where the River Thames might be crossed. Two of these villas are very close to the wood. One at Bix is under a mile away.[3] Another pair of villas in what is now the village of Harpsden lay only two miles to the south. A few pieces of good Roman jewellery have recently been unearthed between Harpsden and Bix by people using metal detectors.

A large complex of buildings, including a high-status villa, has been re-excavated at Yewden, about three miles to the east of the wood (close to my 'snail walk'), and located at a former ford across the Thames; this was evidently a site of considerable strategic importance. A wealth of archaeological evidence uncovered there indicates an occupation spanning three centuries. The most extraordinary, but disturbing, discovery was a large number of skeletal remains of what are termed 'perinatals' – infants who died after thirty-eight to forty weeks' gestation. They are regarded as infanticides.[4] One interpretation of this grisly archive is that the babies are the bones of unwanted offspring originating from a brothel that supplied 'R&R' for off-duty soldiery. There is other evidence from Yewden attesting to the presence of military men who had served the Empire abroad, such as scarab talismans from Egypt.

However this history is eventually decoded, it is clear that our region of the Chiltern Hills was no longer part of any kind of wildwood in Romano-British times. It is inconceivable that those who lived at Harpsden were unaware of their near neighbours at Bix. Surely it is not mere speculation to envisage these people visiting one another to exchange gossip and goods; that is what our species does on Sunday afternoons. The Roman buildings in Harpsden, which are located in a valley, were probably built over newly felled forest. I like to think that the high ground at Bix remained open from Iron Age days, but I have no particular reason to support this supposition, other than the propensity for idleness that is another characteristic of many of our kind. If a widely held view that villas were built over the sites of preceding farms is correct, then the underlying landscape was comprised of fields rather than woods, and the familiar usage probably continued. The summing-up in my court of speculation is that Grim's Dyke Wood had probably not become established in Romano-British times.

Empires decay, but rarely in grand style. When the era of the Roman Chilterns came to an end around AD 410 there was no single cataclysm. Local Britons did not immediately give up all the trappings of civilised Roman life, and pass over a threshold labelled 'Dark Ages' to emerge 650 years later into medieval society. However, by the end of the fifth century and the beginning of the sixth the old order had disappeared. The population had declined after the prosperous centuries. Even coinage – always the archaeologist's best evidence – was abandoned for a while in favour of barter and theft. Christianity had been spread from Rome in the second century, but now polytheistic paganism returned to England, as invading tribes moved in across the Channel from Germany.

The earliest such people to settle around the Chiltern Hills had been mercenaries brought in from northern Europe at the beginning of the fifth century to help defend what remained of the Roman state centred on St Albans (*Verulamium*), and there is archaeological evidence for subsequent Saxon villages in the Vale of Aylesbury and at several places along the Thames. In these early, chaotic times the Saxons avoided the Chiltern highlands for more than 150 years; the hills became what has been described as a 'British reserve'.[5] Our wood was part of the area that held out against the invaders. A number of historians have considered that this redoubt required some kind of centralised political organisation, a last gasp of Rome from St Albans, but it is more likely that the incomers left the area well alone, in the control of tough local warlords. I imagine a warrior stalking through Lambridge on his way to spy out the land from the top of the scarp slope, his ideas of the old Roman days already refracted through several generations of desperate circumstances. Was a golden age still remembered around the campfire? Or did the crude necessities of survival erase shared traditions, rather like barrows obliterated after centuries of the plough? After 571, contact with other relict British areas was cut off by a more organised Saxon assault; the high Chilterns became an enclave. In the plains and valleys beyond the hills the local populations were absorbing new language and new ways. The English tongue was evolving. The British in the hills could not hold out for much longer.

The Chiltern Hills were finally settled by the Saxons in a piecemeal fashion after about AD 650. With the passage of another two hundred years the pattern of the countryside we still see today had been established. This is the period to which we must look for the beginnings of Lambridge Wood. In the British Library is a late-seventh-century document called *The Tribal Hideage* that uses the term 'Chiltern' (*Ciltern saetna*) for the first time, so even the identity of the chalk upland itself can be traced to this period. England was becoming organised again.

Our own area was on the margins of the emerging powers of Mercia and the West Saxons. The settlement of Bensington lay at the foot of the Chiltern Hills a little distance to the west, and was a frontier of great importance – more prominent than Oxford in middle Saxon times. According to *The Anglo-Saxon Chronicle*, Cuthwulf captured it from the Britons in 571. Today, Benson (as it has become) is principally known as the place that regularly records the lowest temperatures in England during cold snaps, although it does still have a warrior presence, in the guise of a modern air force base. In Saxon times, Benson and the surrounding area of the Thames Valley was a focus for both earthly power and spiritual regeneration. Just a few miles upriver, at Dorchester-upon-Thames, St Birinus (d.650) set out to convert the heathens of Wessex to Christianity. His abbey became a site of pilgrimage for centuries – sadly, his supposed relics were probably bogus.

The Mercian King Offa defeated the West Saxon King Cynewulf of Wessex near Bensington in 779, so there could be change at the top.

At the bottom of the social heap it is likely that the remaining Britons were enslaved, or maybe employed as swineherds in the woods. And this mid-Saxon period may well have been when the woods returned to something like their aboriginal state. During their confinement within an enclave, the enduring Britons likely followed upon the early patterns of farming that had been prosecuted in Roman and even Iron Age times. When they were flushed at last from their Chiltern haven and put to work elsewhere, parts of the high land that were not imme-diately useful reverted to scrub, and eventually to woodland. Scrub appears very quickly on fallow land; I have observed fields becoming impassable brush after only a decade if blackthorn, hawthorn and bramble are allowed to have their way. The return of high canopy might take eighty years, to judge from the trees in our own wood. At this stage, shaded woodland floors would soon naturally clear them-selves of all but specially adapted flora. The ancestor of Grim's Dyke Wood established itself on the feather-edge of Saxon civilisation, and thus began its own trajectory through history. The wood is as old as, or even older than, the great Anglo-Saxon poem *Beowulf*, proving that the endurance of art can be tracked by the life of trees; the perpet-ual regeneration of forest parallels the timelessness of literature.

Saxon lords parcelled up their estates in ways that organised the countryside into slices that still exist today. Bensington was 'taken by the king' after 571. It comprised a royal vill – a large estate, with land enough to support parties of nobles if need be. By later Saxon times Bensington's compass included a generous stretch of country running all across our Chiltern territory, from the settlement itself as far as the River Thames near what would become Henley-on-Thames much later. Its outline may even have followed the 'old ditch' of Grim's Dyke, as a convenient, pre-existing line scribed on the ground. Our wood would have been no more than a tiny fragment of the vill. Royal pleas-ure would have required rides for hunting, supplies for feasting, lodges for leisure. Nonetheless, we know that people lived within its bounds. A burial site at Bix has been unearthed to reveal two skele-tons buried clutching coins in their hands, fees for a passage to a better world. The coins belong to the reign of Burgred of Mercia

(852–74), and numismatists recognise, with their wonderful blend of scholarship and history, tokens minted specifically by moneyer Heahwulf in about 865.

For administrative purposes the shire was an important Saxon division, and one only has to hear spats between Yorkshiremen and Lancastrians today to understand how shire identity has embedded itself in regional pride. Administrative convenience has become a variety of tribal fealty: many of our visceral loyalties may at root be Saxon. A shire was administered through a sheriff, the modern connotations of which I do not need to explain. Less resonant today are the 'Hundreds'. This division of a shire was based on the notional resources needed to support a hundred families (for one family, the measurement of a 'hide' persisted into medieval times). In our part of Oxfordshire the Chiltern Hundreds comprise a series of local territorial divisions. Henley-on-Thames – and our wood – is a part of the Binfield Hundred. The adjacent Hundreds are subdivisions of that extensive tract of land embraced by the great southward loop on the course of the River Thames as it negotiates the Goring Gap, and were linked to royal Benson – as 'Bynsington-land' in a charter of 996.[6] One particular feature of the southern Chiltern Hundreds is that they have river frontage. They embrace different kinds of land with particular agricultural and pastoral possibilities: rich floodplain soils closer to the Thames, chalky slopes, and woodland with clearings on the high Chiltern plateau. They comprise broad linear strips across the physiography, providing nearly everything that could be required for a farming community.

After the conversion of Anglo-Saxon England to Christianity, parishes followed similar lines, subdividing the Hundreds into parochial strips almost radially about the course of the River Thames,[7] and continuing in similar fashion further north with slivers at right angles to the Chiltern scarp. Each parish was supplied with land that ensured near self-sufficiency; their distinctive shape always spanned a useful range of natural environments. The better ground grew wheat and barley, which also provided the material for thatching; downland and cleared areas supported domesticated animals; woodland supplied

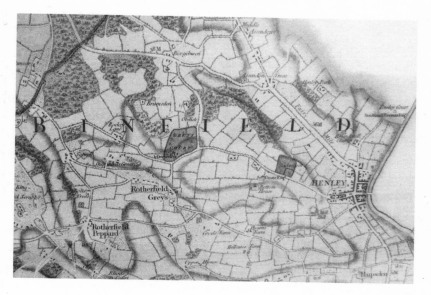

Detail of Richard Davis's map of Oxfordshire, 1797,
showing the Binfield Hundred.

fuel, acorns to fatten pigs, and furnished the raw material to make
huts and halls, bowls and spoons. We shall see how this pattern of
mixed farming lasted for more than a thousand years, and that its
legacy still influences the countryside today. The southern Chiltern
Hills comprise a 'fossil landscape' where the past can be read from the
disposition of woods and hedges, field boundaries and churches, as
much as from excavation or old archives. Maybe this is not an inappropriate place for a palaeontologist like me to come to terms with the
vagaries of human history.

The Anglo-Saxon centuries gave us place names – linguistic fossils,
if you like. The ancient parish in which our wood lies is Rotherfield
Greys. *Rother* is an Anglo-Saxon word for oxen; *feld* is often a cleared
patch of ground. The name should have applied to open land grazed
by cattle. 'Greys' was appended after the Norman Conquest to signify
it as the property of our local lord of the manor, Robert de Grey. The
long, dry valleys so typical of the Chiltern Hills were indicated by the
suffix '-*den*' or '-*don*'. A typical example is Assendon, lying at the end

of the Fair Mile just below our wood, where Cecil Roberts recorded a handful of spelling variations. Harpsden, the excavation site for Roman villas, lies in another, similar valley to the south; Hambleden is to the north. The names are more than labels. The 'rother' tag reminds us that oxen were the working beasts of Saxon times; a pair of these powerful animals was employed to pull the plough and coulter to turn the soil for sowing. The name alone conjures a picture of a farm labourer in a simple shift steering his single-bladed plough across an acre or two of clay-with-flints, whistling encouragement to his team in the winter sunshine as his oxen trudge ahead.

Just over the Chiltern scarp the name of the tiny but perfectly preserved medieval hamlet of Swyncombe records forever that the late Saxon Lord Wigod of Wallingford was wont to hunt wild boar in its sharply bounded valley (the second part of the name is even older – *cwm*, a Celtic dell).[8] However, there is a caution to be attached to old names: land use may have changed; wild boar has disappeared. Turville, on the far side of the Assendon Valley, derives from *thyrre feld*, meaning dry, open country. Now it is largely wooded, but older cottages are still scattered around the edges of a large common area, even though it is much overgrown. Once upon a time it may indeed have been open ground.

Nor are there any Viking names in Oxfordshire, like those commonly found in Yorkshire and eastern parts of England long subject to Danelaw. In the ninth century King Alfred the Great of Wessex had a strongly fortified *burgh* built at Wallingford by the River Thames to keep the threat of Viking pillage at bay. The huge trenches defining his military base are still there, and the site can even be seen from the top of our nearest Chiltern scarp. It was a great work, which has been estimated to have taken ten thousand man hours to dig. In 1002 King Aethelred ordered the slaughter of all the Danes in Oxford, in what is termed the St Brice's Day massacre. Revenge raids by Norsemen on Wallingford led by Sweyn Forkbeard early in the eleventh century must have passed close by our wood as his fearsome crew rowed upstream along the River Thames. By now the hills were spread with trees: oak, beech, ash and lime. The invaders would have sped past this thickly covered and unknown high ground, leaving the forest to the pale ghosts of Romans and the 'horned gods of the coven'.

Ash and after

Earlier generations of Vikings were pagans who venerated the great ash, Yggdrasil, as the World Tree.[9] The gods assemble beneath it, and its branches extend to the heavens, while three delving roots reach to the deep well of wisdom, to the spring that feeds all waters, and to the place where the Norns spin out human fate. Humans instinctively embrace trees as symbols. In nearly all religions the tree of life is a metaphor for commonality of descent. Roots extend to a dark underworld, branches to celestial paradise. The tree of Jesse validates the lineage of Christ. The fruit of the tree of Good and Evil brings sin into the world, and with it the possibility of salvation. Trees are always at the heart of the matter. In art, the tree lends itself to decorative simplification, or can become as complex as a Qum Persian carpet in which every twiglet is bedecked with scarlet plums and persimmons fringed by a border richer than fruitcake. Nor can science escape the attraction of the tree as metaphor or shorthand. The latest computer programs that explore the coded messages of DNA still print out trees diagrammatically to represent new evolutionary relationships. Numbers and names dangle off these trees the way armorial crests illuminate medieval genealogical histories. Every exciting fossil discovery made in Africa by anthropologists is invariably described as 'redrawing the human family tree' – a redraft that has happened more than a dozen times during my lifetime. Trees can clarify or confuse, enlighten or darken, delight or dismay, but they flourish eternally in the crowded forest of the human psyche.

In our wood, ash (*Fraxinus excelsior*) always seems the airiest of the trees, and not at all like deep, dark Yggdrasil. Its canopy is the last to unfurl in spring, and even fully clothed its divided leaves let in more light than do crowns of beech or oak, so that some flowers flourish beneath its benevolent guardianship. I love its Latin species name – *excelsior*! onward and upward! – so apposite for a tree that soars up straight when it is in competition with its neighbours. Ash seems to have a more secure instinct for the truly vertical than beech. Its bark always has a yellow tint to it, and becomes longitudinally fissured at maturity, oddly resembling wrinkled reptilian skin. Dead ash branches retain their bark for some time, and nearly always become studded with black hemispheres the size of golf balls belonging to a fungus called 'King Alfred's cakes' (*Daldinia concentrica*), recalling an incineration for which King Alfred is better known than for fending off Vikings at Wallingford.

Ash is easy to recognise in winter, with its distal branches that dip and then curve up into a rise, like a complex candelabra. Its knobby twigs with sharp black scars like tiny hoofs are not like those of any other dormant tree. In our wood the ash flowers are over before the leaves emerge in May. The male flowers are little more than bundles of crimson whiskery stamens, and the females on the same twig longer, greener, but hardly more conspicuous. When the leaves emerge at last at the ends of the shoots the danger of late frost – to which they are susceptible – is over; each leaf has four or five pairs of willow-like leaflets along the petiole, and always a single leaflet at the tip. By August, bundles of the fruits are beginning to mature, deep

Fast-growing trees: ash (left) and cherry bark compared.

green now, but on their way to bleaching out to compose pale-brownish, dangling bunches of 'keys' the length of matchsticks. Each bladed 'key' has a seed at its base, and when released from its tether in the autumn will twiddle round as it falls like a tiny helicopter (a monocopter perhaps?) to spread away from its parent to a new site. As a strategy for propagation it is effective, as my crowded hordes of young saplings prove. I will have to start thinning them out; and all these came from only four goodly-sized adult ash trees jostling for their place in the canopy.

John Evelyn appreciated ash in *Sylva* as 'so useful and profitable ... next to the oak, that every prudent Lord of a Manor should employ one acre of ground with Ash to every twenty acres of other land: since in as many years it would be worth more than the land itself'. Saxon farmer or feudal woodward alike would have appreciated the qualities of the timber. Naturally strong and elastic, it suits tools that come under pressure, like rakes and harrows, or the axles of carts. All our garden tools in my childhood – forks, spades, hoes – had ash handles, and coupled with Sheffield steel blades they put the cheap and tawdry imitations on offer in today's DIY superstores to shame. They did not snap off at the provocation of a deep rose root or recalcitrant horse-radish. I have to lurk around garage sales to find their like today. Ash also makes all the turned parts – stretchers and bows, spindles and legs – of the traditional Windsor chair, a mighty construction of true endurance. Our own example (plus elm seat) has seen out a century of bottoms, some heavy to start with, and even heavier after a good Sunday lunch.

Ash also supplies the very best firewood. Logs cleave cleanly and burn without complaint. A rhyme by Anonymous explains:

> Beechwood fires are bright and clear
> If the logs are kept a year;
> Oaken logs burn steadily
> If the wood be old and dry;
> But ash dry or ash green
> Makes a fire fit for a queen.

I have also entered wild cherry from our wood into the burning competition. Freshly felled it is impossible, just too full of juice, but after a year's drying it burns cheerfully enough in our hearth with a gentle sweet fragrance, except that a flaming log occasionally spits out gobbets of incendiary shards designed to burn down houses. Ash still wins comfortably.

There *is* one way to make ash immortal, like Yggdrasil. If ash is coppiced it regenerates happily, and will continue to do so indefinitely. A tree has to have attained a certain maturity before its root system is large enough to stand the trauma of the removal of its main trunk. Ash survives this treatment well, and around the low stump left behind after felling will throw up a circle of new shoots in the following year, which will transmute into vigorous young trunks when the old root system plumbs into new shoots, and pumps up fresh growth. A coppice 'stool' results if this procedure is repeated. In East Anglia I have seen coppice stools yards across that are reputed to be eight hundred years old. A crop of standing wood is taken as often as needed – for fork handles or axles or whatever – as long as the tree can survive well. The most vigorous growth in ash happens in the first sixty years, which provides an approximate upper limit for the crop. Maybe twenty years is typical. This is a sustainable harvest as long as the woodsman has the judgement to know the right moment to cull the young trees, and the skill to coppice the tree again without killing it. If everything is done just right, natural regeneration defies time.

Hazel is our own coppice tree. We have too few ash trees to risk an experiment in old coppicing techniques with them. Hazels (*Corylus avellana*) also regenerate wonderfully well. There are seven or eight mature hazels in Grim's Dyke Wood, and all of them have been neglected. Hazels regrow as understory trees if the beeches around them are felled periodically, gifting light. It is likely that our old hazels have been there as long as any beech, though you would not guess it to look at them. They make up a bundle composed of a dense cluster of straight trunks of various thicknesses growing from a common base to twenty-five feet or so. The largest trunks are rather too thick

The interior of the wood in March, before the leaves on the beech trees have unfurled, and while the sky is fully visible.

Transformation: the same view in April, with fresh foliage capturing the sunlight and bluebells carpeting the woodland floor in the distance.

Among the mass of English bluebells one rare white example is a natural variant. Many garden cultivars may have originated from such infrequent 'sports'.

The lesser celandine appears in early spring to catch the sunlight before the canopy closes over. By midsummer this humble plant will have disappeared.

A beech seedling sprouts from the woodland floor. The seed leaves differ markedly from those of the grown tree. Empty beech cupules lie to the right.

Inconspicuous flowers of the holly tree in spring. These are the female flowers with the rudiment of the berry at the centre – the male flowers are found on separate trees.

Wild cherry (gean) in its vernal glory, in flower before the beech leaves behind it have woken up. *Inset:* The flowers are white or the most delicate shade of pink.

Lonny van Ryswyck's experiments with natural materials in the wood. *Above:* Clay and different pigments produce subtly shaded tiles. *Below:* Flints of variable purity can be fired to produce distinct shades of glass. A green ingot was included in the collection.

A plant without chlorophyll, the flowers of the yellow bird's nest (*Monotropa*) rise directly from the ground. The special relationship of this remarkable plant with fungi and beech roots has stimulated new research.

The original painting of Britain's rarest plant, the ghost orchid, *Epipogium aphyllum*, which was discovered in the wood in the twentieth century (National Museum of Wales). This is another remarkable plant that lacks chlorophyll.

Summer nights spent in the wood with Andrew Padmore reveal more than 150 species of moths that are attracted to the light trap. After photography, moths can be released without harm. This Pale Tussock Moth has characteristically furry-looking legs at rest.

The writer struggles with an identification handbook in the light of the moth trap, guided by Andrew. There are endless variations in the exquisite designs of these insects.

The name of the Blood Vein Moth hardly requires explanation. The delicately feathery antennae are the moth's principal sensory equipment. The caterpillars of this moth feed on herbs.

The Purple Thorn Moth is a chunky species with serrated wings. Its caterpillars are among the many species that enjoy browsing on beech leaves.

The Satin Beauty Moth is a perfect example of cryptic colouration, invisible during the daytime against tree bark. The larva will feed on yew.

Butterflies are attracted to the clearings in the wood to bask in the sunlight, like this speckled wood perched on a bramble leaf.

The peacock butterfly flashes its prominent 'eyes' when disturbed, but is cryptically coloured beneath.

A silver-washed fritillary butterfly gorges on bramble nectar; a large and uncommon butterfly, and star of the lepidopteran parade.

The restless comma butterfly has a ragged-looking hind edge to its wings which distinguishes it from all others in the wood.

The red kite is the sentinel of the skies over the wood. Its mewing cry is often heard even when it cannot be seen.

A brown long-eared bat is a specialised predator upon the many moths that throng in the woodland at night.

This little field vole seemed unfazed by our attention after being trapped, and scuttled back gratefully into the wild.

Our rarest mammal, the hazel dormouse, spends much of its time in a torpid state, waking up to feast on fruit and flowers.

The golden hoard of late Iron Age coins found inside a hollow flint close to the wood. At this time the ground was probably clear of trees.

After much searching below the leaf litter these small truffles (*Elaphomyces*) were finally a sufficient reward for a persistent mycologist.

A selection of snails from the wood. All of them are small and thin-shelled, reflecting a shortage of lime for shell construction in this habitat.

The cherry-picker lifting the writer up into the summer canopy of the beech trees. The splayed legs are rather like those of a spider.

This large black terrestrial spider *Coelotes terrestris* is confined to the southern half of England, and was found under our log pile.

Slime mould (*Lycogala epidendrum*) forming pink balls on damp, rotten wood. At this stage their interiors are soft and creamy.

Crab spider *Diaea dorsata* lurking on a holly leaf where its green tints provide an appropriate disguise for a small predator.

A rare crane fly (*Cercophora*) with spectacular antennae, which distinguish it from its many relatives.

to be embraced by my hands. They carry a thin, light-brown glistening bark, with a natural sheen, really no more than a subtle gloss. I have counted thirty-five to fifty annual rings in polished cut sections of the largest trunks, so these trees were following upon the beech trees that now mostly surround them. Like all trees they reach for the light. My clumps probably should have been coppiced several decades ago, but it is not too late.

I selected the most vigorous of our hazel trees to coppice last winter. I had been told that deer find the fresh shoots of regenerating hazels irresistible, so I restrained from cutting right back to the ground, although this does produce the best regrowth. Instead, I chose waist-height. It was an easy task to cut off the small trunks, compared with felling a mature beech tree. The younger hazel poles were long and straight and provided excellent supports for climbing beans. Traditional country uses for such poles were in making the uprights for woven fencing, or stays for the laying of hedges, or spars for thatching. Hazel laths were favoured for producing a supporting lattice for wattle-and-daub walls that 'filled in' between structural oak beams in many vernacular cottages. Finally, hazelnuts are a thoroughly acceptable crop if they can be harvested ahead of greedy squirrels. In sum, the humble hazel is a useful tree by any measure. *The New Sylva* relates that it is known to support 230 invertebrate species, so it is clearly a major asset for biodiversity in any wood.

I began to wonder if I had been too cruel to my first coppice tree, for after being trimmed low, the stool appeared quite dead. For several months the bare stumps remained stubbornly inert. I had decked trimmings from the upper parts of the branches to make a kind of tent over the stool to discourage roe deer that fancied a snack from any regenerating shoots. At last, tiny reddish knobs pushed out of the shiny surface of the bark, a kind of eruption. They soon morphed into shoots. At their growing tips a mass of red hairs emerged along the petioles of the leaves as they began to develop – no other tree in the wood has twigs like this. Then finally big, jagged-edged, point-tipped leaves unfurled fully – the most generously proportioned leaves we have. Now, in August, it is clear that strong new shoots will push towards the light and another crop of coppice is on the way. Deer browsing proved no problem after all.

Like a well-balanced person, the most serviceable woodland has different virtues that rise to different demands. This is not the result of any landowner making a conscious management plan; rather, a diverse range of needs leads naturally enough to an optimal way of exploiting what a wood has to offer. Coppice products are useful, but mature trees are equally necessary to supply timber, and as an item for trade. The eventual outcome is an ancient combination which has been called 'coppice with standards' – a staggered crop rotation, if you will, in which nothing is wasted. The kinds of trees that still grow side by side in Grim's Dyke Wood are an ancient memory of these woodlands past. Hazels might well have been harvested once every fifteen years, with the standards overtopping them growing as singletons, and felled more rarely depending on the type of tree; oak would have been harvested less often than faster-growing beech and ash. The standards also furnished a yearly crop of beechnuts and acorns that served as an ideal food for fattening pigs – rights for this pannage could even be sold off as a separate contract. When areas were cleared, light flooded into the wood, and a succession of woodland flowers appeared, inviting a host of insects in their train, and thence songbirds of many voices. A rich ecological merry-go-round spun onward through centuries. The dreary pine and larch plantations introduced in the twentieth century as a 'cash crop' seem like an insult to time itself.

Clay and after

Lonny van Ryswyck has come to the wood with spades and buckets and strong plastic bags. We have found a place at the edge of the large clearing that seems suitable. Brambles are a nuisance hereabouts. From their established woody stocks new leading shoots have arched over and down and taken root, eventually to interlock in a low-level jungle that will shade out all the weaker herbs. These shoots can be yards long, and are fully protected by thorns. I lash about me with my pick to uproot the invaders. The new blackberry vines pull up easily enough, revealing bunches of starkly white roots wherever they have touched down. A space is soon cleared. Lonny digs down into the orange-brown clay-with-flints, to bring up great gobs of the sticky stuff. The goo is filed in plastic bags to be taken back to the Netherlands, there to be baked into tiles. She pronounces that the clay is 'so beautiful'. Lonny's studio has worked on the polder clays of her native region, rendering in her furnace every farm into its unique ceramic signature, manufacturing a subtly different tile for each property, some deeply red or orange, others with greenish or greyish tints, according to the precise chemical composition of the local material. In this way, she has produced a wall of tiles that encapsulates a whole landscape. So these few bagfuls from our own morsel of Oxfordshire will provide the material for our woods' own special earthen identity. When the small reddish tiles arrive they take pride of place in the collection.

There is a long tradition of brickmaking in our area. Just beyond Lambridge Wood towards Greys Court the name of Brickfield Farm leaves little room for speculation. Very old bricks are not standardised. Neat bricks, smaller than we are used to, make an outbuilding at Greys Court housing the donkey wheel that once brought up water from a well hewn deep down into the Chiltern chalk. They display deep, warm, almost crimson-red tones. Transport was the most important limiting cost in medieval times when buying in all kinds of goods; if raw materials were available locally, then they would be employed to save haulage. There were good local clays. Sand from pits in Lambridge Wood was combined with lime, burned from local chalk using charcoal from the woods, to provide the ingredients for mortar. So all the ingredients for brick buildings were readily supplied from within bounds of the Greys estate. John Hill has discovered the circular base of the old kiln a short distance south-west of Lambridge Wood.

Just up the road on the high ground in Nettlebed, less than three miles from our wood, brickmaking became an industry. A huge, bottled-shaped brick kiln located uncomfortably in a modern residential close is all that remains of centuries of this craft. Nettlebed

The Nettlebed brick kiln.

bricks were said to be the first of their kind made south of the River Humber since the Romans departed. The Stonors were lords of the manor of Nettlebed during the medieval period. Their incomparable family papers record that Thomas Stonor began to add to the manor house at Stonor Park in 1415. During 1416–17 Thomas purchased no fewer than 200,000 'brykes' and employed 'lez Flemyngges' [Flemish people] to build the chapel at the house. These artisans had brought over the necessary skills from the continent and settled down as brickmakers near Nettlebed. £40 was paid to Michael Warwick for the contract to supply the bricks, but another £15 had to be added to transport the load by cart a mere three miles to Stonor. Two hundred and fifty years later the omniscient Dr Robert Plot commented on the toughness of the Nettlebed product, attributing it to a peculiar virtue of the clay. Coppiced woodland doubtless kept the kiln ovens burning all the while.

Bricks from later firings gave local houses a particular signature, since from the eighteenth century brick sizes were roughly standardised. These 'Nettlebed blues' have warmly red sides and blue-grey ends. They are frequently employed in walls to impart a chequerboard effect by alternating stretchers and headers. Houses enhanced in this fashion can be recognised all around Wallingford and Henley-on-Thames, and give many vernacular buildings in this small part of Oxfordshire a delightful but unobtrusive regional identity.

By a curious nomenclatural coincidence the local big house in Nettlebed, Joyce Grove, was occupied in the early twentieth century by a banker, Robert Fleming, who had nothing to do with bricks, but was the grandfather of Ian Fleming, creator of James Bond, a brand as enduring as brickwork; moreover, 'Flemish Bond' is the bricklaying style favoured by those using Nettlebed blues.[10] In 1894 a sale catalogue for the disposal of Stonor lands described Nettlebed as having 'an inexhaustible reserve of clay', but inexhaustible or not, no Nettlebed bricks were made after the 1930s. The former clay pits have now

Sale notice for clay and sand rights, Nettlebed.

almost entirely reverted to secondary woodland. Bricks today usually originate from vast clay pits near Peterborough.

The most readily available and ubiquitous local building stone is flint. It is virtually indestructible, and very hard, but it has problems. Flint does not bind well to mortar. Many flints are not solid, like the stone eggs Dr Plot described. They can be roughly shaped – or knapped – but this is laborious work. A common building style throughout chalk country combines the deficiencies of flint with the virtues of brick, the latter being used to make corners, and to frame doors and windows, while flints of varying shapeliness set in mortar comprise the walls around and between. The best flints are newly dug from the chalk, but respectable ones brought up by the plough will suffice. Traditional brick-and-flint cottages are common in Chiltern country, and are usually charming rather than grand, the former houses of farm labourers and not gentry. Thatched roofs that a Saxon would have readily recognised were once common, but sadly are becoming rarer as fires and insurance premiums take their toll.

Flints were used in grander style in castle walls, as they were in the oldest parts of our wood's manor, Greys Court. Such walls have to be thick to do their job, so masses of flints were jigsawed together in

Kneeler in St Botolph's church, Wigod's hunting area.

120

mortar with any other stony materials that came to hand. Medieval flint churches are also the rule, but here the quoins are made of stone rather than brick. At Swyncombe, tucked into the secret valley where Wigod bagged boars, the diminutive and ancient flint church of St Botolph has limestone blocks on every corner and surrounding the doorways. These stout blocks are not local – they come from Normandy. It was easier to bring them from Caen in the north of the Anglo-Norman kingdom than to import them from around Bath, for example. A special factor here is that until 1404 Swyncombe was part of the important abbey of Bec in Normandy. A similar style of making ancient churches is repeatedly found throughout this part of Oxfordshire. Limestone fonts and stoups were probably all part of the same deal. The walls, however, are from down the road and all around the fields: a mass of rubbly flints, grey and speckled, rough as old oak bark. The overall effect is a building wonderfully at ease with the landscape in which it sits. Its origins may lie still deeper, for the way the flints at the base of the walls are laid in herringbone fashion suggests Saxon handiwork. All the parishes in the manor of Bensington had Saxon forebears, including our own, Rotherfield Greys. The Normans soon replaced the original wooden churches or modest flint constructions they found there. However, like the employment of woodlands, in this 'fossil landscape' almost everything roots back to Saxon times. A thousand years passes easily in the land of rothers and swyne.

Flint-knapping

James Dilley comes to the wood to look for Mesolithic flints, but he fails to find any worked implements on our patch. However, he does find some good-quality flints to show off his skill in knapping. Most of the flints in the wood have a kind of white or brown crust on them. It is the interior that is important. The darker and more uniform it is, the more useful it will be. Flint is hard, but also brittle. If hit in exactly the right way a flake will split off it, with a curved fracture surface. If you have ever broken a thick piece of glass you will recognise this conchoidal fracture, and flint is effectively nothing more than silica glass. Before attempting any work with flint it is as well to wear some eye protection to be on the safe side, and to place a thick cloth over the thighs to prevent any sharp shreds going the wrong way.

In the wood, a suitable stone with which to strike the flint is at hand – one of the rounded liver-coloured sandstones that are scattered over the forest floor. Once a good-looking flint has been found, a good hard tap with the stone detaches a flake or two. Repeated taps in the same direction may even produce some long strips that might make a scraper. The flake often already has a very sharp side, so it is easy to appreciate how a useful tool can be easily acquired. Making anything more sophisticated requires more time and much more skill. A smaller stone can be used to tap off delicate flakes along a cutting edge, for example, and if worked from both sides something like a knife can result. An additional tool that James produces, a club made from deer antler, is surprisingly effective at removing small flakes. The antler is far softer than flint, but sharp percussion exploits its brittleness. An hour or two spent in this pursuit is guaranteed to increase your admiration for 'Stone Age' technologists.

6

September

Gold!

A singular golden brilliance floods Grim's Dyke: September sunlight
thrown at a low angle through tree trunks that act as a series of pillars
or bars. The leaves still provide a full canopy, so the generous light
penetrates only in patches that are landlocked pools of incandescence.
A hanging mist is a hint of autumn in waiting. As I walk through the
darkness of the holly glade the far prospect is one of golden paradise
seen from the underworld. Then I find my own gold: no Iron Age
hoard, but just as much shining treasure. My own chanterelles! They
glow like trapped sunbeams by the path that runs through the centre
of the wood. Small, bright-yellow mushrooms shaped like funnels
push through the beech litter over a patch three yards across. They are
the colour of egg yolks, or the stamens of saffron. They have a bright-
ness that seems almost suspiciously enthusiastic among last year's dull
leaves. I am momentarily torn between leaving the chanterelles
(*Cantharellus cibarius*) to decorate the pathside or scooping them up
and carrying them home as golden booty. They have already shed
millions of spores, so their reproductive mission is complete: the prize
is rightly mine. They go straight into my basket.

The evening light is now fading fast, but still illuminates white
porcelain mushrooms sprouting from a pile of felled beech logs that
our neighbour has stacked close to the barn. If the chanterelles
presented as unrealistically yellow, these clusters of mushrooms glow
too white to be true, as if they were some other-worldly china replica

of the real thing. Their caps drip glutinous slime that picks up glints from the sun's last rays. Every bit of the porcelain fungus (*Oudemansiella mucida*) is eerily white – the gills beneath the cap, the ring that decorates the stem. It is said that the caps are edible if the gluten is washed away, but I am not tempted. I am reminded of scientific research on this particular mushroom. The porcelain fungus defends its domain within the dead beech trunk on which it feeds by making chemicals (strobilurins) that repel fungal competitors – a kind of molecular face-off. These same chemicals have proved unusually effective as ecologically friendly fungicides; they are used everywhere to fight off harmful moulds that might otherwise damage crops, and without poisoning benign creatures. There are some golden rewards that are hidden from sunlight.

When I climb into the better light provided by the car I discover another gift from the wood. Enchanter's nightshade seeds have stuck in their dozens to my trousers, looking like so many little greenish insects. The least spectacular of our woodland flowers, creeping unheeded under brambles and where bluebells enjoyed their brief glory, this modest species probably finished flowering in July. These tiny, paired seeds are the last part of its life cycle. Richard Mabey[1] explains that its rather wonderful name is a translation from its botanical Latin equivalent, *Circaea lutetiana*, which has to do with the Greek enchantress Circe. In the sixteenth century Parisian botanists were of the opinion that this plant was the source of the potion that the witch used to turn Ulysses' crew into pigs. *Lutetia* being a classical name for the forerunner of Paris, this becomes 'Circe's poison of the Parisians', a glamorous name for such a simple herb (in fact, it may well be non-poisonous).

However, its seeds are something of which a magician would not be ashamed. Under my hand lens I see that the green, swollen clubs are coated in a mass of tiny white hairs, and that each hair is hooked at its tip. After the unspectacular flowers have done their job, the seeds hang pendulous at the tips of pin-like stalks, and must easily detach to hook a ride on any passing animal. As far as they are concerned, my socks and trousers are just the furry transport required to spread them to a new patch. For a moment I wonder at the perfection of the design, and feel a sneaking sympathy for those who believe in a Designer; the attention to detail does indeed seem so awesome. Everything works at such a small scale, even down to the molecules that can scare off competitive moulds. Is this not an exquisite collaboration greater than we can comprehend? A divine plan? Yet I know that this 'explanation' is a spell spun by a deceiving enchantress. The wood is a complex weave of adaptations. Every feature of every organism has been tested by natural selection. An enchanter's nightshade seed that failed to produce those most perfectly 'designed' hooks would succeed less well in propagating itself. A porcelain fungus whose chemical shield failed would soon be selected out of the population. The enchantress would have me leave behind the very rational faculties that reveal the wonder and pleasure of understanding nature. For the scientist the analytical process does not diminish the splendour of what he or she sees. Every detail added is an extra stanza added to a great epic poem, one that is never complete, nor yet ever tedious in its particulars.

I brush off the seeds of the enchanter's nightshade rather carefully before I drive home. If the plant establishes itself in my garden, it will become a pernicious weed.

Manor and town

Our wood became part of one of the manors of Rotherfield shortly after the Norman Conquest in 1066. The estate is recorded in the great tally of the Domesday Book twenty years later as held by Anketil de Grey, and various de Grey descendants held court there for more than three hundred years, during which time their name became appended to the parish. Another prominent family, the Pippards, managed a similar trick further to the west, thereby labelling the village of Rotherfield Peppard as it appears on maps today.[2] Domesday also tells us that the estate was valued at £5 to the lord, that it supported twenty households (twelve villeins and eight bordars), had seven ploughlands, twelve acres of meadow and four-by-four furlongs of woodland. It was already presenting a mixed landscape, rather as it does today. There never was a village of Rotherfield Greys in the cosy, clustered-round-the-green sense; it always comprised scattered farmhouses and workers' cottages, all looking towards Greys Court at the centre.

Today, only curtain walls and three towers remain of the medieval castle that the de Greys built, but their flinty bulk still impresses where they are aligned opposite the later Tudor mansion as tattered but defiant relics of serious times. One of the ancient towers has been incorporated into a seventeenth-century dower house, and must command one of the most enviable views in the south of England, over flawless rolling countryside. By contrast, a medieval traveller glimpsing the original castle from across the same valley would have found it quite intimidating.

It is likely that Greys Court was surrounded by wood pasture, at least after 1200; an open park studded with mature trees where deer could thrive alongside other livestock. 'Ploughland' was used for arable crops, and is certain to have been on the good farmland lying north of the big house that is still partly employed for that purpose. The area occupied by woodland was actually rather less than it is now – I estimate about 135 acres. There are, of course, no maps of the estate from these early days, so there will always be some uncertainty about land use, but it is reasonable to suppose that the woodland lay furthest from the manor towards the edges of the property, beyond the worked fields, and more or less where it is today.

The parish of Rotherfield Greys was one of those long, thin strips of ground that had been established in Saxon times. At least five miles long and never more than a mile wide, and often much less, it extended eastwards to include its own Thames river frontage. This would become more important as trade grew beyond the immediate area, but from the first it provided a private fishery for the lord's use. The parish included a range of woodland, grassland, and productive agricultural land sufficient to support a slowly increasing population, all the way from the high 'waste' ground at Highmoor to the rich meadows along the Thameside. Lack of water in the more elevated ground led to the construction of several ponds, and eventually to the excavation of a few deep wells.

The feudal system demanded duty of work from the tenants, and as the Greys' manor was kept 'in demesne' through much of the Middle Ages the peasantry were bound in thrall to the lord of the manor for generation after generation. They worked the fields, tended the animals, and respected the trees. These anonymous serfs were more familiar with our wood than anyone else. The Luttrell Psalter (c.1330) allows a glimpse of them: dressed in plain, pleated coarsely woven tunics, with chausses clothing their legs and a simple hood, or chaperon, covering the head. Toil and religion dominated their lives. Only saints' days permitted them brief respite from labour. In the wood, they coppiced and felled as necessary, making bundles of faggots for fuel, posts for fencing, repairing tools with ash handles, maybe secretly smuggling home a piece of beech wood to craft a simple bowl. Some tenants had rights to gather brash for fuel (known as *woodbote*), and there was a requirement on the part of the lord to release wood to repair dwellings (*housebote*), but almost everything in the lives of tied men was specified by a list of obligations and duties. They would have been appalled to see how much fallen wood just rots away today. We do not know how our peasantry fared during the terrible civil war period known as the Anarchy (1135–54), but as confrontations between the rival claimants to the throne, Stephen and Matilda, were played out in siege and slaughter at Wallingford (1153) so close by, it is unlikely that they would have escaped the general suffering. Hobbes's famous phrase on the effects of warfare could have been coined specifically for this dismal period: 'the life of man, solitary, poore, nasty, brutish, and short'. Even the gentry suffered.

By the early thirteenth century the de Grey name was celebrated. Walter de Grey was one of the most powerful clerics in the land during the reigns of King John and Henry III. John appointed him Chancellor in 1214, and then Bishop of Worcester; he was at the King's side when Magna Carta was signed on 15 June 1215. He was subsequently elected Archbishop of York, which made him a very rich man. When Henry succeeded to the throne Walter de Grey was still an important figure, even effectively running the kingdom for a brief period in 1242 while the King engaged in a disastrous adventure at Poitou in an attempt to recover lost French territories. Walter de Grey's embalmed body was reverentially laid to rest in the south transept of York Minster in 1255, marked by a mass for the passing of a great man. A succession of de Greys followed with title to the manor at Greys Court. The first Baron Grey de Rotherfield was John de Grey (1300–59), one of the original members of the Most Noble Order of the Garter; he was Lord Steward to the Household of Edward III. The beautiful brass commemorating Lord Robert de Grey (d.1387) in full armour in Rotherfield Greys' church records the late stage of the dynasty.

Whatever great business went on in court or church associated with the big house, the quotidian business of the wood continued with billhook and cleaver, coppice and twine, almost unremarked by their lordships. Trees grew as reputations waxed, trees were felled as reputations waned. The transient vanities of worldly advancement or disappointment left only a few tree rings to measure the passage of ambition in the harder currency of arboreal growth. The trees in the wood read only the truth of the seasons, the gifts of rain, the hardships of drought.

Even as the medieval prosperity of Greys Court increased, just over two miles to the east a transformation was in train that would redirect the history of the southern Chiltern Hills, our wood included. Henley-on-Thames was created. The town did not grow chaotically from urban predecessors like many others in England, but was planned and laid out from the beginning. It was built on royal land, a morsel of the ancient manor of Bensington (today's Benson) with a long Saxon pedigree. The exact date of its design is disputed,[3] but it is likely that some major work was undertaken in the 1170s, during the reign of Henry II – a monarch known for creating planned towns elsewhere. Henley's first bridge, at least partly built of stone, was constructed across the River Thames at about the same time. Its founding arch can still be seen in the cellar of the old waterside inn, The Angel on the Bridge, where it now rather ignominiously shelters beer kegs. St

Mary's church lies at the edge of town close to the end of today's bridge, and dates originally from 1204, although much modified in the ensuing centuries. Next to the bridge were a series of wharves that would secure the future of Henley.

The rest of the town plan comprised a rough quadrilateral centred on the same Market Place that is there today. The developed land was divided into long and narrow burgage plots. These stretched behind the shopfronts that faced on to the street, and accommodated all manner of workshops for trades that would contribute to the further growth of the town. The inquisitive walker along the east side of Bell Street can still nose around between modern premises belonging to estate agents or coffee houses or whatever and discover the burgage plots ranged behind, for the original design has endured with remarkably little change. Modern developers scratch their heads to think what they can do with a template designed in the Middle Ages; while they do so we can be grateful that a 'fossil townscape' is preserved in appropriate counterpart to the ancient landscape that survives in the hills beyond town.

Ancient entrances in Henley reveal long
burgage plots behind the houses.

Extraordinary remnants of these early days can be discovered among the shops and pubs that line the oldest streets in Henley. The modern passer-by would be hard put to recognise them for what they are, although arrays of later Tudor beams are on brazen display here and there. Many more buildings hide their antiquity as if it were a guilty secret. The original constructions were so robust that subsequent builders left old oak struts and beams in place, and simply accommodated them within new designs – but they can still be smelled out. In Bell Street good evidence of a fine medieval hall was discovered, with massive oak beams that a dendrochronologist has dated to 1325, when Edward II was on the throne of England. Blackened beams even attest to former fires that burned openly in long-lost hearths. One splendid, somewhat later medieval building is tucked away behind the big church. Beautifully moulded oak timbers on its exterior suggest that the building was intended to impress; the first floor provided showrooms and warehouses for the leading Henley merchants John Elmes and John Deven in the mid-1400s. It

The Chantry House, one of Henley's finest medieval buildings.

was close to the wharf on its eastern side, where stores of grain, malt and wool adjoined a bustling waterfront. The Chantry House is evidence for a new entrepreneurial class, confident and skilled at spotting business opportunities, and shows how much had changed in local society since early feudal times. The planned town of Henley-on-Thames had become a thriving success.

River trade was driven by the needs of London. During the early medieval period an increasingly prosperous and populous capital city[4] had to be fed and comforted. Wheat and other cereals were in constant demand, as was fish, fresh or salted; wood was the only fuel for winter heating; charcoal was required for smelting and for making gunpowder; leather goods and wool clothed burghers and common-ers alike. Such commodities arrived at London wharves by boat along the River Thames, as they would have done for centuries; but Henley was created in just the right place at the right time to become a major entrepôt. The river below Henley was readily navigable by large, flat-bottomed vessels known as 'shouts' that could negotiate the shallow stretches of water on their way to the capital.

The Thames upstream from Henley was a lighterman's nightmare. The great loop in the river past Reading was not only difficult to navi-gate, but was beset with fish traps and water-driven flour mills. Critical stretches of the river could only be negotiated by way of flash locks – inefficient predecessors of the pound locks employed today – which worked by staunching the flow of the stream with wooden gates. Downstream traffic was floated through when the gates were opened, while upstream traffic had to be winched through the lock at exactly the right moment. Millers resented the diversion of their power source, while local manorial lords had an interest both in the mills and in collecting fees from passing boatmen. There were only four flash locks between Henley and London, but more than twenty upstream on the way to Oxford. The economic case for Henley was made: goods were best loaded and unloaded at wharves alongside the new town. London merchants established premises there, no doubt including the fine halls lurking beneath the veneer of the next six centuries. Property deeds for the town between 1280 and 1350 explicitly list cornmongers and fishmongers among the tradesmen. Granaries were built to safeguard cereal crops brought in from the surrounding hills before export to London, where the merchants owned yet other wharves.

As for trade in the other direction, goods unloaded at Henley were taken over the Chiltern Hills by packhorse and cart to Wallingford, bypassing that slow, expensive and inconvenient loop in the river, before continuing on to Oxford. Several of these routes had originally been established by drovers centuries earlier. Wagons would have trundled laboriously past our wood as they made their way back and forth over the hills. Our coppicer would have heard the cries of the hurrying muleteers as he bent to his work, and the ploughman, alerted by the groan of axle against shaft, would have paused briefly to watch loads of merchandise lumbering gracelessly along the track past Greys Court. Henley became the conduit through which all commerce flowed.

The effect of this hubbub on the manor at Rotherfield Greys can well be imagined. A self-sufficient rural economy gave way to one with a closer eye on the market. Wheat became a commodity, and the

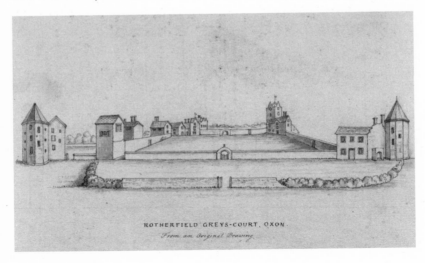

ROTHERFIELD GREYS-COURT, OXON.
From an Original Drawing.

In this eighteenth-century drawing, Greys Court retains
many of its medieval features.

products of our wood were another. The de Greys were worldly-wise enough to have a stake in the changing economy; our small piece of the Chilterns became part of the wider world for the first time. If the free market favoured wheat there would be an incentive to grub up woods – using every scrap of beech, no doubt, in the process – and take the ground into arable use, a practice known as assarting. As the medieval population grew, the rewards of arable farming were more tempting. Agricultural methods improved: big pits excavated into the chalk in the lower part of Lambridge Wood may well have supplied lime to 'sweeten' fields underlain by clay-with-flints, to increase its productivity for cereal crops. There was even a section of waterfront belonging to Rotherfield Greys right next to the southern Henley wharf, a bonus derived from the oddly elongate Saxon shape of the parish. There was nothing to stop our estate trading wood and grain independently. The old road through the parish runs from the river up Friday Street (now part of Henley), thence up Greys Hill to leave town at a point where the police regularly catch motorists breaking the 30 mph speed limit in their understandable hurry to reach St Nicholas church a mile away. I don't doubt that horse-drawn carts struggled up the same slopes half a millennium ago when there was no speed limit to worry about, but ruts aplenty to slow them down.

Oak

The growth of Henley required timber. Oak trees provided the only spars, beams and crucks from which substantial frame buildings could be constructed. Oak comes with a set of natural adjectives, all of them somehow chunky and comforting: words like 'stalwart', 'trusty', 'stout-hearted' and 'sturdy'. Many Englishmen flatter themselves that their characters partake of the same qualities assigned to oak, although species of the genus *Quercus* are generously spread about the world, and stalwart, trusty chums are rare enough in any currency. John Evelyn in *Sylva* lauded oak's qualities as 'of all timber products hitherto known, the most universally useful and strong; for though some trees be harder … yet we find them more fragil, and not as well qualified to support great incumbencies and weights, nor is there any timber more lasting'. Oaks mean business, they are not fly-by-nights, nor do they fight shy of any incumbencies put upon them.

Anxiety about an insufficient supply of mature oaks for the navy was what prompted Evelyn's famous survey in the first place. Grim's Dyke Wood has only two mature oaks now, and a third planted by our family with some ceremony, but there are further good trees scattered through Lambridge Wood, and every reason to suppose that they were more numerous in the past than they are now. Oaks were frequently chosen as standard trees with coppice beneath them, and the dominance of beech today may not have been ever thus if oaks had been selectively harvested in the past.

Trees were taken out of the Oxfordshire woods to make those great medieval halls, and we know that this first happened before the fourteenth century. The frame was always laid out before it was erected, under the instructions of master carpenters. Their scribbled work marks occasionally survive. Although nothing in Henley compares with the masterpieces built in London, such as the roof of Westminster Hall,[5] crafting these durable frames was a skill requiring exquisite judgement and knowledge of materials. 'Green' wood was preferred for structural work, because when it contracted during its subsequent seasoning this tended to seal and tighten the whole frame. Additional oak was used for doors and windows, and sawn planks for floors in the best houses. Walls were often no more than 'filling in' with wattle-and-daub, or with bricks laid herringbone fashion in later dwellings. The real work was always done by oak.

It is impossible to imagine life without oak timber in the pre-industrial age. The stalwart friend of the carpenter was also raw material for the cooper's barrels. The wheelwright needed oak spokes, and shipbuilding was unthinkable without oak to furnish both strength in the hold and luxury in the captain's quarters. Oak was the employable tree, an indispensable resource, imbued with its special qualities of reliability and endurance. Oak probably reached its *literary* apotheosis after its golden age had passed. I call as evidence a later version of 'Heart of Oak' presented by Reverend Rylance on 16 April 1809:

> When Alfred, our King, drove the Dane from this land,
>   He planted an oak with his own royal hand;
> And he pray'd for Heaven's blessing to hallow the tree,
>   As a sceptre for England, the queen of the sea.
>   Heart of oak are our ships,
>   Heart of oak are our men ...[6]

The sturdy oak has morphed to become the most vital part of the internal anatomy of the true Englishman. King Alfred the Great would likely have planted his supposed oak at his redoubt in Wallingford, just a few miles from our wood, in curiously psychic anticipation of a naval supremacy that would follow five hundred years later – although admittedly it is true that half a millennium is not an exceptional age for an oak that has been left to mature through its whole life cycle. Today, truly ancient oaks are mostly found scattered in what remains of medieval wood pasture around great houses,

Oak (left) and beech bark compared.

where they have been forgotten by time, or used to mark estate boundaries. I think of Chaucer's ancient trees from *The Knight's Tale*:

With knotty, knarry, bareyne trees olde,
Of stubbes sharpe and hidouse to biholde.

Jackie and I have discovered some hollowed-out oaken Methuselahs along the old highways over the Chiltern Hills leading towards Oxford, and around a pond at a high point there where sheep and cattle must have been watered on their way to market; but there is nothing like them in working woods like ours. Oak is too useful to be allowed to grow old gracefully.

At the Museum of Rural Life in Reading, black-and-white archive films show oak craftsmen at work. A local cooper demonstrates deft precision born of long apprenticeship and years of practice in turning a bundle of oak staves and a few iron hoops into a barrel or tun that can last for many years. He embodies an everyday miracle combining sympathy for wood with hand–eye coordination, and his sureness of touch leaves a do-it-yourself duffer like me awash with awe. In old Henley he would have been much employed in making kegs to transport fish, and later in manufacturing beer barrels and wine casks; with the original implication of the Latin *manus* and *fecit* – 'made by hand'. Turned beech was used for the bungs, so our commonest tree had its own small part to play. The fact that Cooper is still such a common surname proves the former ubiquity of the trade. It is as well that the best wines and whiskies still enjoy residence in oak barrels, or the cooper's craft may well have vanished by now. That of the cartwright (not an uncommon surname either) or wheelwright is more endangered, but probably not as much as the ghost orchid. A wooden wheel requires a whole parade of the trees from our wood: wych elm for the hub, ash for the felloes – the curved sections making the outer wooden rim of the wheel – and reliable oak for the spokes. I wonder whether our wood could ever have been under contract to a wheelwright.

Those who compare oak with hearts or barques or courage always seem to forget the tanning. Medieval serfs, freemen and grandees marched on leather boots; jerkins made of leather covered their torsos; merchants carried their money in leather pouches; even common soldiers were mostly protected by leather armour. But the journey from an animal skin to leather was complex, tanning being the most crucial part of it; and tanning required oak bark. Tanneries stank. They were placed at the edge of town. In Henley, Friday Street was (formerly) at the boundary with Rotherfield Greys parish, and the poor souls of our parish received the pong from the tannery. After scraping off the bad meat and bristles, which must have been malodorous enough, skins were steeped for a long time in a ghastly potion involving chicken or even dog faeces, with urine as an optional extra. This immersion induced chemical changes in the skins that took the process to the next stage; they were then doused with oak-bark liquor, which contains tannins to complete the rather marvellous transformation of tough hide into supple leather.

It seems astonishing to me that this process was discovered at all, but I have inspected a *urinarium* in Pompeii where human effluent was stored for the tanners' use a thousand years before Henley was founded. It was a profitable business, too, so tanners could indeed become stinking rich; one of the finest double-fronted houses on Friday Street is the tanner's house. The Stonor papers[7] tell us that the bark was bought on the standing tree, which was stripped off on felling: the timber went one way, the bark another – and more than three dozen trees were needed for an average tanner's year. The very best bark was regularly harvested from oak coppice on a twenty-year cycle. I can find no overgrown coppice stools in Lambridge Wood, but I can be perfectly sure that when an oak standard was felled its bark did not go to waste. It went to leather.

I love to see oak branches and twigs in winter. They are all dark elbows and angles against the sky, quite at odds with the elegant or sweeping lines of beech and ash. I love to feel oak's thick and coarsely ribbed grey bark, so emphatic compared with the gentle surface of beech. Our species is the pedunculate, or English oak (*Quercus robur*) – heart of oak itself – and the finer of our two trees is probably about eighty years old, the same age as several of our more mature beeches.

Its trunk rises sheer and straight among its beechen neighbours. In spring, oak leaves appear later than on nearly all the beech trees, but they are all so far up that I only receive occasional messages from on high to keep up with the oak's progress. Oak catkins are little more than strings of stamens, and when they fall to the ground in May they are a slightly sorry sight, like threadbare mouse tails. Wind-blown oak leaves tumble down from time to time. Some of them display spangle galls on their undersides – pink-flushed discs that look like tiny press-studs applied to the leaf surface. They are produced from the leaf after stimulation by chemicals secreted by a minute gall wasp with a long name: *Neuroterus quercusbaccinum*. Something like five hundred invertebrate species are supported by oak trees, more than any other forest tree, so the canopy must be thronging with insect life if only I could reach it. Tannins are supposed to offer the tree some protection by making its leaves more unpalatable, but plenty of animals have evidently found a way to cope with this distraction.

This year is a good one for the acorns that now lie scattered all over the ground under the tree. It is a 'mast year' – something that happens about one year in five. Long ago, wild boar would have feasted on the fallen acorns, and medieval swineherds would have fattened up their pigs on the back of their brief profusion, but many years have passed since rights of pannage were taken up for the wood.[8] Nowadays, the abundant yield in a mast year is under-utilised. Many of the acorns still sit in their scaly cupules, which hold them as snugly as eggcups supporting eggs. I take out one acorn, which leaves me holding the cupule, and the stem on which it sits – the peduncle that gives the English oak its Latin name. As a child I often put this little contraption into my mouth as a pretend pipe. It feels so small now. This oak could see a dozen of my lifetimes and still have many good mast years ahead of it. I put that profound inevitability into my pretend pipe and pretend to smoke it.

Some of the acorns are still green, but most have faded to tan; but here is something that is not an acorn, though not so dissimilar in size. Perfectly spherical, and polished brown, this marble gall was produced as the tree reacted to another tiny wasp, *Andricus kollari*. This specimen is for the collection. I can see the wasp's tiny exit holes from the safe house full of food that it persuaded the tree to make for it. I have never seen the wasp itself, which is not surprising, since it is only two millimetres long. Oaks support more species of gall wasps than any other tree, and each species stimulates a gall with a highly specific shape. It is a wonder how the tree 'knows' to do this, and I am not clear how one shape rather than another fits in with the life of a particular wasp species. Dryads were Greek nymphs as specific as a gall wasp to the oak tree; maybe they would have been able to tell me the answer.

Oak trees became an important part of the economy from medieval times. Within the Greys Court estate the old woodward would have known trees as individuals. He would have understood when to leave well alone and when to harvest. Now the estate has passed from the hands of lords of the manor – the woods sold off, the old house and grounds run by the National Trust and managed for visitors. The stewards of what is now an Oxfordshire 'day out' look to the danger to the public from falling trees rather than to profits from tanning or pannage. The oak, however, remains in charge. The National Trust logo is a stylised oak sprig: four leaves, one acorn, a few cupules. What else? After all, oak is somehow 'national' (*'heart of oak ...'*); and 'trust' – like honesty, durability and stalwartness – is another of those words that attaches to oak like a gall.

Truffles

I am crouched under a beech tree a little way from where its main roots are anchored into the thin soil. In my right hand I have a tiny rake bought from a toyshop. I have noticed little diggings into the beech litter around this particular tree. I set about scuffling under the surface with my slightly comical implement. In a few moments I have scraped away some of the loose leaves and have got down to something like soil and rotting litter. It is so stony that I have to remove flints as they get in the way, but quite soon I see slightly puffy pinkish rootlets belonging to my beech tree – they branch profusely like little trees themselves. I know that these are roots with mycorrhiza (p.72) on the outside. Beechnuts seem to have hidden themselves perversely in the ground to resemble tiny truffles – after all, they are brown, bean-shaped, and buried. I carry on with a little less enthusiasm than when I started, since I have already failed several times in my search. Then the rake brings up something different: rounded bodies looking like rusty-brown rabbit-droppings, as large as big peas. I have struck lucky at last. Under my hand lens I observe that the little objects have a typically warty surface. After looking for so long I have recovered four truffles in as many minutes.

When truffles are mentioned, most people begin to salivate and put on a mental bib and tucker. They think of the celebrated black *truffe de Périgord*, or of the most expensive foodstuff in the world, the white truffle of Italy, the Château Petrus of the gastronomic milieu, weight for weight more valuable than gold. They seem disappointed when confronted with four rabbit-droppings. Most truffles are modest objects of this kind. There are numerous species, but since all of them develop underground they are unfamiliar even to experienced naturalists. They have to be hunted down. I was taught the routine by Caroline Hobart, doyenne of English trufflers, whose customary post on a fungus foray is underneath a spreading tree with a blanket to sit on, and a small rake to scrape with, and a collecting box to house her finds. It is more like archaeology than mushrooming.

I know the particular truffle species I have just discovered – one of the common ones, *Elaphomyces muricatus*. When cut through, the interior of a truffle usually has the consistency of hard cheese, and has a characteristically marbled appearance, although the general ground colour can vary. When first dug up truffles can resemble potatoes, or cape gooseberries, or snowberries, or rabbit-droppings. In truth, a truffle is not a particular *thing* at all: it is a habit. Let me explain. Several groups of fungi have discovered the advantage of growing to maturity without breaking above ground. Their spores develop within a sack-like fruit body. Most of these species are spread by animal activity – especially grubbers like wild pigs – and to help this process they evolved the most delicious smells. These wondrous odours are fugacious; an old truffle does not smell of all kinds of lip-smacking meaty umami rolled into one, but just rather stale. (Beware all cheap-truffle vendors, for they are selling a half-remembered dream.) The spores pass unharmed through the digestive system of the animal and are spread through the wood.

It has been known for a long time that truffles arose from the two major groups of larger fungi. Some truffles are related to honest-to-goodness 'shop' mushrooms (Basidiomycetes), while others are closer to morels (Ascomycetes) – so they had entirely separate evolutionary origins.[9] All the delicious ones belong to the second group (my species was also an 'asco', but not one of the good ones). This phenomenon of similar form evolving in response to similar adaptations is common in nature, and I am glad to have it demonstrated in the wood – it is often called 'convergence'. What was not fully realised until the molecular age was that the truffle habit had evolved separately even *within* the two major groups. Many different fungi wanted to have a go, as it were, at being a truffle. Sequences obtained from genes extracted from the DNA of the truffles have identified their closest relatives among several, only remotely related, kinds of fungi. This means that rather similar-looking truffles are now included in the same classification alongside very different-looking mushrooms, proving again that evolution can cheat common sense. All truffles seem to be associated with tree roots in a mycorrhizal relationship, so they donate useful nutrients to the tree in return for sugars derived from photosynthesis. By maturing underground (*hypogeal habit*) truffles avoid the perils of spore dispersal in air, where fruiting bodies are always at risk of desiccation. The spores go straight into an efficient dispersal mechanism – otherwise known as a digestive tract. It works very well for the fungus, and equally well for the trees that are partnered by the fungus, which receive a home delivery of the right collaborator gift-wrapped in a parcel of dung. It is the kind of adaptive package upon which evolution really likes to get to work – hence the recurrent truffle habit.

I wish I could say that I had discovered truffles in the wood by spotting the tiny truffle flies (*Suillia pallida*) that hang around the location of buried mycological treasure like bloodhounds round a corpse in a shallow grave. Trufflers have claimed to be able to master this entomological trick, and I almost believe them. The summer truffle (*Tuber aestivum*) is our native edible species, and I would dearly love to find it in the wood, with or without the help of flies. My lesser discovery was prompted not by any insect, but by a series of blatant scrapes down into the litter. I wondered about a badger on the prowl, but the size was probably more appropriate for squirrel activity. Nothing with olfactory organs can resist the lure of a truffle, even if it means coming down from the safety of the trees.

The fly man cometh

I am ignorant when it comes to flies, and not just truffle flies; but then, there are an astounding number of flies to be ignorant *about*. I need help. I worked alongside Dick Vane-Wright for several decades at the Natural History Museum – or, to be more accurate, his office was about half a mile away from mine in that great warren of a building, but we both knew its secret ways. Dick has an unquenchable enthusiasm for crane flies, and he is making one of his visits to the wood in search of them. Flies have two wings to fly with rather than four – hence the name of the insect order to which they belong: Diptera (true flies: 'two wings' in Greek). The second pair of wings in flies is modified to become distinctive balancing organs. Fly experts are known as dipterists.

Dick will have to stand in for several dipterists, each with their own favourite kinds of flies, who came to Grim's Dyke Wood to use the cherry-picker. The antics of collectors in acquiring their samples are not dissimilar. Crane flies ('daddy longlegs' to some) will lead the show, because they are the largest of the fly persuasion. With their spindly legs and long, skinny wings they are a familiar sight just this month, bumbling and blundering around meadows, shedding limbs. Dick himself has a certain leggy presence: tall and gangly, with beard and straggling wispy white locks, he bears an increasing resemblance to Gandalf. It is a warm, slightly humid day – a good one for flies. Dick arrives with a net, and not a particularly discreet one; in fact, it resembles a rather voluminous pair of white Victorian bloomers mounted on a stick. To bring up the crane flies he prances up and down in the vegetation by the path, at the same time making broad sweeps and lunges with the net. A dog-walker passes by, and the dog looks even more astonished than his owner at the goings-on. He (the dog) cannot bring himself to bark. For the smaller insects, there is a short pause to suck out the catch from the base of the net into a collecting jar using a 'pooter', but the larger crane flies are popped directly into a killer jar. They have to be taken away and studied under a binocular microscope to be accurately identified.

So what is the problem? A crane fly is just a crane fly, surely? Absolutely not! The challenge is that there are more kinds of crane flies than there are species of birds. Six hundred and eighty-eight genera of crane flies have been recognised, before you pass on to the 15,391 named species; and remember that previously unnamed new species are still being discovered every year. Just over three hundred species are known from Great Britain. There are actually very few people who can identify crane flies with confidence. With the exception of most of the Lepidoptera (butterflies and moths), this is the problem with many kinds of insects, and is the reason why the list of species from Grim's Dyke Wood is still incomplete. The crane flies are just one family group within the true flies, and there are over a hundred more families.

For most of them, I await the right expert. However, the crane flies have one extraordinary advantage: Charles P. Alexander (1889–1981). He is not as well known as John James Audubon, maestro of the birds, let alone Charles Darwin, the naturalist's naturalist. In the annals of crane flies, however, he reigns supreme. Alexander scientifically named and described no fewer than 11,278 species, one for almost every day of his career. Two-thirds of all crane flies were named by him, and he published 1,054 scientific papers, which may be a world record. I don't imagine he had much of a social life. There are invertebrate species waiting to be discovered in every part of the world, from tropical rainforest to deep oceans, and probably even in our wood. Scientists disagree on just what proportion of all species we already know, but even the most conservative taxonomist would accept that we currently recognise under half of the world's biodiversity.[10] To those who say that the task of collecting and describing this vast inventory of unknown species is impossible, simply too overwhelming, there is one unanswerable riposte: Charles P. Alexander.

The rituals of identification become second nature to an entomologist. One specialist might focus on colour patterns, another on mouthparts, according to the type of insect concerned, but I believe that almost all authorities are interested in genitalia. On reflection, almost everyone I know is interested in genitalia. The wings of flying insects such as crane flies are supported by a series of struts known as veins, and are further divided into compartments called 'cells'. These display great constancy within a group, and vary from one group to another, which is meat and drink to an entomologist. The sexes of crane flies differ. The male of one of our wood species has gorgeously fluffy antennae, for example, in comparison with a less endowed female. Colour spots appear on the wings of other species; legs can get stripy. As for the genitalia at the tip of the long abdomen, they comprise a complex, whiskery sort of arrangement that you would have to be an entomologist to know and love. They are of crucial importance to the insect – not least to ensure that the right partner is mated – which makes them equally of interest to Dick Vane-Wright.

A common crane fly recovered from the wood, *Tipula oleracea*, was named by the great Linnaeus in 1758, and so is twinned with our beech trees in scientific history. It could even be claimed that this species, by historical precedence, is the daddy of all daddy longlegs. Its larva is an unlovely brown grub known as a leatherjacket, which causes serious damage to roots of garden plants and is loathed by horticulturalists, although much appreciated by hungry birds. Dick provided me with a set of detailed notes on the other fifteen species of crane flies recovered from the wood. His descriptions have an esoteric poetic quality when he describes 'the slightly infumed area of the more posterior wing membrane', or a 'maculated wing pattern', or maybe 'exquisite bipectinate antennae'. This is the love of a naturalist for his very own bugs.

Yet I had been bugged by a question myself: I wondered how the wood could support so many different species of crane flies. Now I know. Their leatherjackets have specialised preferences, and mature at different times of year. *Tricyphona immaculata* likes the soggy conifer woodpile at the edge of the track; *Rhiphidia maculata* prefers rotting beech elsewhere in the wood, and emerges after *Ormosia lineata* in the same habitat. A common species with pretty, spotted wings, *Limonia nubeculosa*, is more of a generalist, or an 'opportunistic facultative xylophage', as Dick succinctly, if technically, puts it; *Cheilotrichia cinerascens* was collected around holly; *Erioptera lutea* lives in soil. The larvae of *Ula mixta* are fungivores, probably feeding on decaying bracket fungi. This particular species was only named in 1983, and might have been considered a rarity until recently, but it is more likely that there was just nobody around to recognise it for what it was. The largest crane fly of all was *Tipula maxima*, with mottled wings, whose larva lives in damp soil – possibly from the permanently moist area along the track. However, the most spectacular species is coloured up all black and yellow along the body like a thin wasp: *Ctenophora flaveolata*. Its comb-like antennae resemble the artificial eyelashes of a 1950s *femme fatale*. It is not at all common. 'A nice beast to find,' says Dick. Its last recorded occurrence in Oxfordshire in 1993 was thought worthy of a note in the *British Journal of Entomology and Natural History*; it was from the Warburg Reserve, two and a half miles north of us: 'a species of ancient beech woods', remarked its discoverer.

Details, details, you might protest. I reply that the delight, as well as the devil, is in the details. To an animal of small size, particularly an insect in which the larva hides away discreetly to feed and grow, our wood is a *potpourri* of opportunities, quite a wonderland of niches. 'Biodiversity' as a word sounds rather dull and a bit abstract. Played out on the ground it is something else: the difference between the numbered title of a symphony and its glorious complexity unwrapped in a concert hall. Every rotting log is a small world. The underside of a leaf is a realm to a greenfly; a crack in the bark of a beech tree is a capacious and secret hideout. They all fit together in a jigsaw that remakes its own pieces month by month. Crane flies emerge only briefly into the common democracy of the air to meet their mates and continue the species; otherwise, they are sequestered in secret places, biding their time … A late sunbeam cuts across the clearing, and picks out all the flying insects as dancing motes. Each little life is not much more than a pinhead of brilliance. Some bob up and down together; a mating dance, I suppose. Others move purposefully and then vanish from the light, seeking something, smelling something, following a precise instinct to a precise niche. Even if I had the scientific names of them all it would only be like having the notes on a page, not the symphony. A species inventory is only the beginning. Every species will have its own biography, its special requirements, and its curious secrets. Dick tells me that the early stages of some of the crane flies in our wood are still not known. How much more is there still to find out about the hundreds of insects adrift in this aerial confetti?

Chanterelle potatoes

This is what happened to those chanterelles: a simple way to get the most out of their flavour. Potatoes absorb the juices released from the mushrooms as they cook. Like all wild food, quantities vary according to how many chanterelles you have been lucky enough to find, but they should be about half the weight of the potatoes. I lightly cook a small sliced onion or shallot in olive oil until it is soft. Then I add sliced, lightly cooked floury potatoes and brown them a little, and next the chanterelles, coarsely sliced if they are big ones. I cover the pan at this point to help them 'sweat' over low heat, and particularly in damp weather they will release a lot of fluid. When this has happened, I remove the cover and allow the liquor to reduce satisfactorily. I do not like the potatoes to wholly dry out – a juicy tackiness is about right. This part takes about eight minutes or so. When served with a couple of rashers of good bacon this dish is all that is needed for supper.

7

October

Nuts

It is raining beechnuts. They fall to the ground with a succession of clicks, like a metallic version of a cloudburst. Beechnuts cover the ground till it is crunchy to walk on. The lower branches still hang on to the husks – or cupules – that enclosed the nuts. Cupules are small, gaping woody purses divided into four lobes, all shiny and smoothly brown on the inside, and coarsely hairy on the outside. They will loiter around on the ground long after the nuts have disappeared. Every nut is a three-sided cylinder, designed so that four nuts can snuggle together within the cupule as they ripen. The crop is abundant this year. Grey squirrels will be the principal beneficiaries, damn them, but then, so will wood mice and many kinds of birds. Wood pigeons will spend the winter scratching around to discover nuts they missed in October.

Not every polished brown nut has a kernel – some are just pretending, and prove to be empty. But when most of the plumper ones are peeled with a fingernail a pale, yellow-brown, faintly three-lobed little kernel can be released without much trouble. It tastes good too, but it would take hours to gather and peel enough nuts for a decent snack. As I kneel to pick up more of them I am distracted by a fallen beech leaf with two tiny, yellowish, hairy towers perched on top of it. They look like microscopic versions of shaggy boots. I am alert to galls by now, and sure enough this is another one, stimulated into growth by a tiny midge called *Hartigiola annulipes*. It is much easier to collect and identify the gall than the insects responsible for it. As with habitual criminals, their *modus operandi* always gives them away.

The Chiltern Society is arriving to clear and reinstate the public footpaths through Lambridge Wood. Over many years people have just wandered around the beech woods, creating their own habitual ways. The Society is there to put them back on track, following the old rights of way. In charge of the operation is Stephen Fox, burly and effective, and a bit bossy, as he needs to be. The team of volunteers is comprised of senior citizens and is a very jolly bunch, fit and game and jokey. Picks and mattocks, saws and strong clippers are all supplied. 'We all have to be a bit nuts to do this,' says one amiable white-haired lady wielding a billhook. Among their number are a retired doctor, and a man who worked for the BBC. The Rights of Way Officer of Oxfordshire County Council is there to see fair play. They chop down small trees blocking the path and grub up brambles obscuring the proper track. Felled holly branches are laid alongside the path both as a way of defining it and to allow their spiny leaves to deter wanderers. 'Unofficial' paths are blocked off with more serious pieces of timber. A robin bounces along behind the party to see what edible tidbits are turned up during the clearing. It is surprising how fast the team progresses through the woodland, all the way up to the end of Grim's Dyke Wood. 'We like to get to the pub in good time,' explains the man formerly from the BBC.

I was worried that determined locals would just continue to follow their old routes, and would resent 'busybodies' telling them what to do. A path along the part of Grim's Dyke that passes through our wood is a case in point: it is not a right of way, but people have been using it for years, partly because this option is drier during the soggy winter months. The correct path, now clearly indicated by the Chiltern Society with a smart new signpost, runs nearly parallel a few yards away; and it can get quite wet. Sure enough, a determined walker soon removed all the branches providing a barrier at the end of the spurious track on the dyke. Jackie and I moved them back again. So began a kind of war of attrition, which continues. We moved some more serious branches from a felled tree to provide a more serious barrier.

By the end of the following week they too had been moved again by our mystery walker. So for the ensuing month we staggered through our wood carrying branches we could barely lift. That would fix it. This time the spars were so heavy that they were swivelled, rather than lifted, to reopen the track for our unknown interloper. I began to understand the psychology of an arms race. I had ideas about moving whole trees that *nobody* could shift, using big machines: the ultimate deterrent. The competition was clearly driving me nuts.

Then I came to my senses. Public footpaths are one of the most precious features of the English countryside. The right of people to walk where their forebears walked for centuries past is an important principle to defend. I have defended it myself in parts of Wales where farmers are cavalier about putting bulls in fields crossed by footpaths and allowing their stiles to collapse, and I have scars to prove it. Our little wood is almost circumscribed by three footpaths. I had already discovered that dog-walkers kept deer under control, so there were advantages to having passers-by. On the other hand, Lambridge Wood is protected as a Site of Special Scientific Interest (SSSI) for its ancient beech woodland.[1] This imposes regulations about what I can and cannot do – for example, I am not allowed to fell a big tree without first getting approval from Natural England and the Forestry Commission. It is illegal for people to come into my wood and dig up bluebells. I have a duty to protect rare species: the Dutchman's pipe growing near Grim's Dyke needs me to care about it, and the dyke itself is an ancient monument that deserves to survive for another thousand years.

We get mad at cyclists who use our footpaths – cycle routes are in a different category from pedestrian rights of way. For their part, some cyclists are cheerily oblivious to the distinction. On one occasion I threw myself in front of a miniature jeep grinding noisily along one of the paths. Tolerance has its limits (the driver backed off with moderate grace). What I really want to do is to alert the joggers and the cyclists to the delights of examining the details of the habitat they pass through so quickly and with so little regard. I want people to stop and read its biography from nature, to enjoy it as I do. Who knows?

Maybe I could even convince the persistent log-shifter to abandon old habits.

Greys Court affairs

The history of our wood and the story of Greys Court are intimately entwined. But even an estate like Rotherfield Greys cannot be fenced off from the outside world, as its game park might be from the privations of the peasantry. The early medieval period was one of increasing population and growth in trade. This accompanied a long phase of benign climate. If it had continued, woods marginal to the estate could well have been grubbed up and taken into arable use. The fourteenth century threw everything into reverse. The Great Famine of 1315–17 saw a succession of wet summers and implacable winters. Cereal crops failed. Seed corn was eaten in desperation. Starvation and the diseases encouraged by it took a huge toll; up to 20 per cent of the English population is thought to have perished. Cannibalism became a common crime. Global climatic change was caused by vast quantities of volcanic dust and gas released into the atmosphere by the catastrophic eruption of Mount Tarawera in New Zealand. Our wood was inexorably linked to events at the other end of the world. In the hard currency of tree girth this led to three years where little heartwood was added, just thin and measly growth rings that dendrochronologists use to recognise as a signature of those desperate years.

The Black Death followed in 1348–49, when bubonic plague swept through the land, respecting neither privilege nor estate. The historian Simon Townley estimates that Henley fared particularly badly because of its close connections with London – up to two-thirds of the population may have died. Afterwards, entrepreneurs moved back quickly, turning tragedy into opportunity. William Woodhall appears in the town records in 1350, was Town Warden two years later, and by his death in 1358 had a trading business stretching through south-east England. Henley kept its reputation as a commercial centre, and by the fifteenth century had enhanced it still further. From the narrow perspective of our wood, years of disaster meant that there was no pressure from population growth and so the forest was safe from clearance (assarting). It continued its useful life, its links with the past unbroken. But by now it should be clear that there were also links that connected our small piece of woodland in Oxfordshire with what was happening in the wider world. These links remain: they bind the whole biosphere together in common climatic cause. Viewed this way, the estate is global.

Generations of de Greys survived the difficult years, based at the manor house, which had by now lost any worthwhile function as a castle. The land was worked on through season after season, sustained by strict routines, but when labour became scarce after the ravages wrought by hunger and disease, statutes were passed to ensure that no advantage accrued to the workers. The interests of lords and merchants were, of course, protected. During a succession of minority heirs to Greys Court between 1399 and 1439 the old buildings became run-down. Later in the fifteenth century the title to the manor passed by way of marriage to the Lovell family and their relatives for several decades. Francis, Lord Lovell, was responsible for breaking the long lineage of manorial rights stretching back to the Norman Conquest. He elected to support Richard III at the Battle of Bosworth in 1485. For this treasonable action his estate was forfeited. This is a summary from an *Inquisition* held on 2 March in Henley-on-Thames in 1514, the fifth year of King Henry VIII's reign: 'Francis Lovell was convicted and attainted by a certain act of Parliament held at Westminster 7th November 1485 … Jasper, Duke of Bedford occupied the manor from the time of the attainder until the time of his death, to wit 21st December 1495.'[2] In 1503 King Henry VII granted Rotherfield Greys to Robert Knollys (pronounced 'Noles'), Gentleman Usher of the King's Chamber. By 1514, Henry VIII required the payment of just a single red rose at midsummer as settlement of the manor upon Robert and his wife, Lettice. The Tudor zenith of Greys Court was to follow.

Robert Knollys' son Sir Francis (?1512–96) was deeply embroiled with Tudor royalty. His sincerely held Protestant views were in tune with the age, except during Mary Tudor's brief Catholic reign (1553–

58), when he was obliged to flee to Frankfurt. He married Katherine Carey in 1545; she was the daughter of Mary Boleyn, and both a first cousin and a good friend of the future Queen Elizabeth. When Francis and Katherine were en route to exile during Mary's ascendancy the Princess Elizabeth had written advising her to 'think the pilgrimage rather a proof of your friends, than a leaving of your country'. Mary Boleyn had been a mistress of Henry VIII before he married her younger sister, Anne. Recent research has offered support to the idea that Katherine was the illegitimate child of Henry, born during his affair with Mary.[3] Mary's story became the basis of Philippa Gregory's 2001 novel *The Other Boleyn Girl*, and the movie of the same name. Whatever her paternity, Katherine was fecund. Francis Knollys recorded details of her fourteen births in a Latin dictionary that has recently come to light. Her offspring line up in alabaster tribute around the splendid Knollys monument in the chapel appended to the parish church at Rotherfield Greys.

The Knollys thrived at the royal court when Elizabeth became Queen. Katherine was Chief Lady of the Queen's Bedchamber. Francis became Gloriana's close adviser, a Privy Councillor, a Vice-Chamberlain, and Treasurer of the Royal Household from 1572 to 1596. He was required to guard Mary, Queen of Scots at Bolton Castle in 1568, which must say something about the confidence Elizabeth placed in him. His Oxfordshire career was as remarkable: joint Lord Lieutenant of Oxfordshire in 1560, and High Steward of the City of Oxford in 1564. None of these appointments were without their perquisites. Portraits show a slightly stiff, confident, serious courtier. Francis remodelled Greys Court to the best standards of the Elizabethan age, demolishing much of what remained of the medieval castle, and creating the fine, triple-gabled house that greets National Trust visitors today. Red bricks made from clay dug from the Greys Court fields built further housing ranges to the west, and a deep well ensured a constant supply of sweet water from the chalk. Greys Court now had all the hallmarks of royal approbation, and the sense of security that comes from wise career decisions. The woods at the edge of the estate were not threatened.

Families do not necessarily follow the course their parents might wish for them. Francis's eldest son Henry had more than a dash of the black sheep about him. In 1575 he became quite the buccaneer with his ship *The Elephant*. Three years later he tried to join Sir Francis Gilbert on his colonising expedition to America with his brother Francis Jr, but instead they set sail early to capture plunder from a famous pirate. The expedition departed without them. By 1581 Henry Knollys was in trouble again for attacking Portuguese ships on behalf of the pretender to the throne in Lisbon, against the explicit orders of Elizabeth. He died in the Netherlands under unknown circumstances. His elder sister Lettice was born at Greys Court on 8 November 1543. When she was seventeen she married Walter Devereux, Second Viscount Hereford (and later Earl of Essex), with whom she had five children. In 1576 her husband died, allegedly of dysentery, in Dublin, where he had spent much of the previous three years. The beautiful Lettice was already embroiled in an affair with Robert Dudley, Earl of Leicester, who happened also to be the favourite of Queen Elizabeth. Lettice and Robert married privately at Wanstead in Essex on 21 September 1578. The Queen was furious. It was twenty years before Elizabeth would allow Lettice into her presence, and she is famously reported to have dubbed her a 'she-wolf'.

None of this seems to have overly concerned the wilful daughter of Greys Court, who outlived all three of her husbands, dying at more than ninety years of age, a remarkable achievement in Tudor times. As for her connection with our woods, it is known that she loved hunting, and it seems plausible that she honed her skills in the chase at her childhood home. I like to think of her stalking through our woods in search of deer, maybe intuiting the techniques of evasion that would serve her well later. She hunted with her younger brother, William, who became much more the son of the father, holding positions at court at the end of the sixteenth century. He was famously vain; indeed, he has been fingered as the model for the character of Malvolio in Shakespeare's *Twelfth Night*. He was responsible for the erection of the exuberant monument to his father in his own parish church. William's effigy with that of his wife sits on top of the elaborate canopy covering Francis and his family, as if he were kneeling before a desk with a book in front of him. The image is oddly touching.

Henley-on-Thames was now booming. In the period 1568–73 one-third of all grain bound for London came through the town's wharves. Most of this was wheat for bread, but local entrepreneurs realised that 'added value' could be released by malting barley in the town, which rapidly became one of Henley's important trades. The arable land and woodland mix that had been part of the character of the Chiltern Hills since Saxon times got a new lease of life. As London grew, the demand for fuel rose commensurately. During the middle part of the sixteenth century the price of wood rose by more than 70 per cent. In 1559 the London mayor and aldermen were allowed to ship in no fewer than six thousand loads of wood stored at Henley and Weybridge. Beech wood was particularly popular for fuel, and no tree in the Chiltern Hills needed to go to waste. Much of the wood was harvested from coppices on a ten-to-fifteen-year cycle to provide billets and bavins, which were standardised wood measures. Billets were three feet four inches long with a circumference of ten inches, and were particularly suitable for burning in the wide hearths of the time, where open spits were the rule. Bavins were three feet long with a circumference of twenty-four inches; we might now describe them as large logs.

By 1543 there was already anxiety about wood supplies, which led to one of the very first pieces of conservation legislation, the Statute of the Woods. It was decreed that young trees were to be left to regenerate after any woodland clearance, and fines could be imposed for grubbing out woodland to extend agricultural land. Beeches grew faster than oak trees, so the good profits to be made from firewood may have prompted changes in forest management that led to the dominance of what we now regard as the typical Chiltern tree. An irregular line of laden carts creaked over rustic tracks along the Assendon Valley from Stonor House, or over the summit from Nettlebed, before gratefully reaching the flat, straight road along the Fair Mile on their way to stack up by the Henley wharves. Except for the 'standards' left for longer-term cropping – oaks for ships' timbers, perhaps – the Tudor coppiced woods would have seemed scrubby compared with the soaring beeches we see now. Whether Lambridge Wood received this treatment is impossible to say for certain, but its proximity to the straightest road into Henley is good reason to suppose that it did. The earliest published map of our area, by Christopher Saxton, dates from 1570, but the portrayal of land use is schematic at best; emparked areas around big houses are shown, and would not have been felled. It is altogether probable that another phase in the life of our wood had begun.

The Fair Mile, from Burn's *History of Henley on Thames*, 1861.

London no longer had a monopoly on the resources derived from the area around Henley. During Tudor times Oxford had begun to shake itself out of its medieval torpor. Six new colleges of the university were founded in the 1500s, including that on the largest site, St John's. Many Oxford colleges held land, including woods, in the Chiltern Hills, and their meticulous written records often provide the best hard evidence pertaining to the use of the natural resources that supported the scholars. Some of these ancient ownerships survive today, and names like 'College Wood' or 'Queen's Wood' on the map around Henley make the connection plain. Even that great impediment to riverine trade, the southerly loop that diverted the River Thames, had begun to yield to determined traffic. The probate inventory of a trader called Thomas West (d.1573/74)[4] proves that at least some commercial river craft were able to make the journey from Henley upstream as far as Culham, south of Oxford, although the final stretch into the city still defied navigation. Notwithstanding these brave essays, it would be many decades before the special role of Henley as a centre of trade diminished. Meanwhile, the southern part of the Chiltern countryside was becoming part of a larger world, just as the Knollys linked Greys Court with the diplomacy and adventures that founded Elizabeth I's global ambitions. It wasn't just volcanic eruptions that started to stitch the world together.

Mushrooms

Eruptions are happening again – all over the wood. Candle-snuff fungus erupts from every old stump. In places it looks as if the ground is sprouting white whiskers; on closer inspection the 'whiskers' are more like little antlers, with the tips of their tines all dusty with white spores.[5] October is the glory time for a fungus-lover. Just the right quantity of rain followed by just the perfect dose of sunshine encourages a frenzied upsurge of mushrooms. They erupt everywhere; pushing out of the leaf litter, decking dead trees, and covering much of the old conifer log pile. Where can I even start? A gallery of shapes: some stout and dumpy, some spindly, some solitary and elegant, some densely clustered. Little conical caps with slender stems seem to be designed especially as fairy parasols. Bright colours defy the fading light, for here some ruby-red caps glow like dropped jewels. Troops of tiny mushrooms too numerous to count all seek to throw their spores into the breeze.

One is special to our Chiltern beech woods: saffron drop bonnet (*Mycena crocata*), eight elegant little mushrooms arising in a line from a fallen log. I break a brittle stem and a spurt of orange juice stains my fingers bright carrot colour. So many kinds of fungi vie for my attention – there must be ten times as many different kinds decorating the leaf litter as there are flowering plants in the same woodland. Many are the regular umbrella mushroom-shape, but some look like clumps of pale-brown upright coral (*Ramaria stricta*) making a miniature terrestrial reef around a tree stump. White puffballs (*Lycoperdon perlatum*) sprout as if the ground itself had inflated a series of miniature party balloons from within the humus. A beech branch is decked with tiers of delicate brackets that are softly hairy above, like cats' ears. Even a dead bramble is ornamented with a rosette of tiny white mushrooms with caps no bigger than my fingernails (*Marasmiellus ramealis*). Mundane twigs are decorated with delicate mycological beauties that might have been crafted by the masters of Meissen porcelain.

The colours on display span the whole palette except blue – though I have even seen azure mushrooms outside the wood. Some mushrooms are white as cartoon ghosts, others as brilliantly red as a slapped cheek, or yellow as egg yolk – but then there are others that are paler, more like clotted cream. There is even a green mushroom (*Russula virescens* – green cracked brittlegill), though it is an unearthly green that has nothing to do with the green of chlorophyll, the tint of photosynthesis. This fungus is coloured by a completely different pigment, cheekily assuming the livery of flowering plants. Every shade of brown that can be imagined is here, from umber to tan, so many tiny things that hardly stand out among the leaves, discreet as if they wish to evade discovery. They are referred to in the trade as LBMs (little brown mushrooms), and a good measure of dedication is required to identify them. Many forayers just pass them by, but they are a precious component of our diversity, so I shall persist with them. There are even black fungi, the colour of death (like *Xylaria polymorpha*, 'dead man's fingers'), but these are not morbid – indeed, they have just sprung to life. Some mushroom caps glisten all smooth and shiny in the autumnal sunlight, while others have a strange gleam when the thick slime that covers them is illuminated. Tiny toadstools spring from single leaves, like a parade of pins. In October the wood is briefly a polychromatic pageant, when the fungi reach their reproductive apogee. It is the mushroom-fancier's brief heaven.

If only mushrooms announced themselves in their true colours. For those interested in eating them only red is a warning, as red mushrooms are, in general, poisonous. Otherwise, the colour of the caps is irrelevant or even misleading. There are dozens of species with white caps like the field mushroom, and some of them are highly poisonous. The most deadly mushroom of them all – the death cap (*Amanita phalloides*) – has yet to be found in our wood (though common nearby) and is a special yellow-green. I suppose I find this shade particularly sinister because I know how lethal this mushroom is; but then, I have eaten another green mushroom with relish. To identify mushrooms, closer observation is needed.

Mushroom-enthusiasts soon learn to look under the cap. Most mushrooms have gills there, but some have a spongy surface instead, composed of many tiny tubes, a feature which identifies the boletes. Several species in the wood include one in which the tubes bruise blue-green as soon as they are touched. For identification purposes the colour of mushroom gills can be more important than that of the cap. Gills can be white or black or deep pink, or various shades of brown, and even change from one colour to another in some species. Since the gills are effectively spore factories, very often the colour of the gills is also similar to that of the spores – though with fungi there are *always* exceptions. The experienced field forayer always notices how the gills are attached to the stem: some types of mushroom have gills that stop short of it, others just reach it, others again have gills that run across the underside of the cap to abut the stem directly. Still others have gills that turn towards the ground to run down the stem, giving the whole fruit body a funnel-like appearance. Some fungi have a veil which envelops the young mushroom. As the cap expands it bursts its covering, which may be left as a bag at the base, or breaks up into patches that are scattered over the cap. The white 'spots' on the archetypal red toadstool, the fly agaric (*Amanita muscaria*), are not spots at all, they are fragments of this universal veil that can be moved with a fingernail. Many more mushrooms have another veil that joins the edge of the young cap to the stem. When the cap expands the remnant is left behind as a ring (annulus) that often hangs down on the stem, or encircles it like a bracelet.

So far the questing mycologist has employed just one of his or her senses: vision. Identifying mushrooms uses *all* the senses, other than hearing – although if I am completely truthful, I should admit that some fungi do actually speak to me. Touch is important, but hard to describe. Cap surfaces of some fungi have the 'feel' of a kid glove, or of velvet. Fingertips respond with the sensitivity of a microscope to details of the cells that make up the surface of the cap. Taste is critical for several identifications, since some very similar-looking fungi betray their identity in the mouth. There is a special way to do the tasting – a tiny piece is nibbled on the end of the tongue long enough to release its characteristic chemicals. In the wood, more than half a dozen species of milk mushrooms (*Lactarius*) have been discovered; these are fleshy fungi that exude characteristic latex when the flesh is broken. To taste it, a tiny drop of milk on the tip of the tongue usually suffices. Some species are as mild as water, some have a distinctly hot taste, and a few would eclipse a vindaloo curry. *Lactarius pyrogalus*, growing under hazel, is one of the last kind, and is found in the wood. Its Latin name means 'fire milk', which says all you need to know.

Here I must confess to a piece of wickedness. I lead fungus forays around our area, and it occasionally happens that one member of the party is particularly irritating. It might be an over-confident and hectoring individual, but the one that tops the bill is the little man who asks of every fungus we find: 'Can you eat it?' while ignoring all the interesting stuff I am attempting to retail about spores, mycorrhiza, rarity, adaptation, etc. The question is particularly exasperating if the mushroom to hand is the size of a daisy, and would require a dozen to make a mouthful. 'I need someone to perform a taste test,' I announce, while holding up a specimen of *Lactarius pyrogalus*. 'Can you eat it?' asks Mr Irritating. 'No, but you can taste it,' I reply. A dab of 'milk' on the tongue and my tormentor is silenced for an hour.

The most difficult sense is smell. Many mushrooms have a distinctive odour, but few people can agree on what it is. A handful of species do spark an easy consensus: one in our wood is the sulphur knight, *Tricholoma sulfureum*, which everyone recognises as smelling of coal tar – it positively reeks of road-mending. *Mycena pura* smells of radishes, and most people recognise the smell, but not without prompting. An allied odour is that of cut potatoes, which is typical of the false death cap, *Amanita citrina*. The list of smells goes on: pears, pencil shavings, drawing ink, and most implausibly of all, 'wet chicken feathers' (*Clitocybe phaeophthalma*). The matter that exercises me is how the original discoverer came across the analogy. How many people carry wet chickens around with them?

I have now identified well over three hundred species of fungi from my small wood. The list is some measure of the diversity of this great biological kingdom, one that is neither animal nor plant. Lists are very dull, like telephone directories, but every name in a directory leads to the biography of a particular individual, which is much more interesting; and so every wild species has its own special story. In previous chapters the stinkhorn, truffle, chanterelle and sulphur polypore have stepped out from the anonymity of the list. I have spent time with many of the three hundred mushrooms: time collecting, time tasting and sniffing, time in my laboratory at home looking at their spores or gills under a microscope to get their identification correct. If I recounted all their biographies there would be nothing else in this book. Fortunately, fungi do fall into categories according to how they earn a living – more like a business directory, if you prefer.

Many fungi break down what plants have made: wood, stems or leaves – deconstructing cellulose and turning it into their food. These are the saprobes. Without them, the forest would soon silt up with branches and leaves, and little nourishment would return to the soil. The wood would become impassable. Some of the commonest October mushrooms among the leaf litter are saprobes, often fruiting in fairy rings: toughshanks (*Gymnopus*) and funnel caps (*Clitocybe*) appear every year without fail.[6] The upright coral (*Ramaria stricta*) grows from buried wood. Nearly twenty different species of delicate and beautiful bonnets (*Mycena*) grow on fallen trunks or twigs, or tiny ones even on single leaves.

A fungus foray in Grim's Dyke Wood.

With fungi, size really doesn't matter: a mushroom the size of a tack is just as interesting as one the size of a soup plate. A few fungus species are modestly-proportioned brownish 'weeds', and are the first to appear along pathsides after summer rainstorms, and as quick to disappear. From dead stumps and sticks erupt tough, bracket-shaped fungi, with pores rather than gills as the site of the spore factories on their undersides; the turkey tail (*Trametes versicolor*) is the most abundant, and for once the common name is a precise description. The banded fans of this small bracket really *do* look like flaunted turkey tails. Then there are small saprobes that specialise in feeding on beech cupules: a miniature mushroom, *Flammulaster carpophila*, and clusters of a white, waxy, disc-like ascomycete (*Phaeohelotium fagineum*) among them. Other tiny species thrive only on grass stems or on fallen bramble leaves. There is a whole universe of rot and decay in the wood, as the right mushroom seeks its favourite place to grow. This is one reason why many millions of spores are produced – so very few of them will alight in exactly the right place at the right time to germinate into a viable feeding hypha. It is a hit-and-miss affair. Because the fungi see their ecological niches so precisely, an ordinary wood can support hundreds of species. One robust, inedible mushroom (*Hebeloma radicosum*) is only associated with mouse middens.

Almost as many fungi in the wood are symbiotic with trees, forming mycorrhiza with their roots, to their mutual benefit (see p.72). They do not have to produce mushrooms every year if the conditions are unfavourable. The most brightly coloured and conspicuous of these mushrooms are brittlegills (*Russula*), with a dozen species in the wood, every shade of white, yellow, green, purple, pink or red, sometimes mixed together. One species, the blackening brittlegill (*Russula nigricans*), changes over several weeks from white to black especially to confuse the bewildered beginner. All brittlegills have a curious texture – the stem snaps in an abrupt fashion when broken, just like a stick of old-fashioned blackboard chalk. This is because the cells that make up the fruit body are minute spheres rather than thread-like, as they are in most mushrooms. The milkcap mushrooms (*Lactarius*) are a related group of species with a more muted colour range. The commonest one in the wood is a sticky-capped, greenish form with a penchant for beech (beech milkcap *Lactarius blennius*). Different *Lactarius* species are associated with oak and with hazel; they are very picky. The wood is not the best place for *Amanita* – the group that includes the familiar 'red-and-white-spotted' fly agaric under which pixies traditionally cavort (*A. muscaria*). I have found it once on our ground. False death cap (*A. citrina*) and blusher (*A. rubescens*) are commonest, and a supposedly rare species (*A. lividopallescens*) comes up regularly. The largest group of mushrooms of them all, the webcaps (*Cortinarius*), sport cobweb-like veils, and are strictly mycorrhizal; several kinds occur in the wood, although they are notoriously hard to identify.

So I will be partial, and select the most precious mycorrhizal mushroom for celebration.[7] A baby cep is the most satisfying thing to find. Usually, slugs despoil the fruit bodies as soon as they appear, so a perfect one is a true joy. The stem is chunky and fattest towards the base, like a big pestle. It is decorated with a fine white network. It sits comfortably on the hand. The cap is hemispherical, often almost exactly so, with a tan-brown tint of something fresh from the oven, and often with a thin white line around the edge. Pores (not gills) are white, or later yellowish, and do not change colour when rubbed. The whole cep has a solid presence, as if it were the product of a ceramicist rather than of nature. I look for signs of maggots: mushroom flies adore this substantial object as much as humans. Where one cep is found, there are often others around. They might be older, more nibbled and less beautiful, but they will not go to waste. Older specimens have shed nearly all their spores, so they can be picked without a second thought. I have seen Italian farmers guard their ceps with a shotgun over their knees; in Italy there are fungus deaths every year. Not from gunshot, however. They are caused by mushroom-lovers in the mountains tumbling to their deaths after reaching out just a little too far over the edge of a precipice for the perfect baby porcini.

Older ceps are excellent for drying. They will keep for a year (or more), and are easily reconstituted by pouring on a little hot water and waiting. The soaking juice is always used in any cookery that follows. The smell of a jar of dried ceps is reputed to cheer up gloomy Russians (the entire population of the former Soviet Union) in the depths of winter, and indeed does have a wonderful wholesomeness, a promise of all good things. Mature ceps should have the tubes removed from under the cap – they can be pushed off with the thumb (they tend to go soggy). This is the time to discard the wormiest bits, usually the lower part of the stem. How much is thrown away is a matter of individual squeamishness. If you examine packets of dried ceps on which good money has been spent you will surely notice little wandering holes where maggots enjoyed their feast. I do not mind a few 'uninvited guests' – they turn to dust in the drying process. I cut up all the remaining mushroom into thin slices, a few millimetres thick; plenty of slices from a big cep, then spread out the slices on newspaper (only a good-quality broadsheet will do). I have under-floor central heating that is ideal to complete the process after I have laid the cep-charged papers on it. An airing cupboard is also fine for the purpose. Damp ceps will soon yield a dark patch on the paper, because mushrooms do hold water, especially in damp weather. In a few days the cep slices will have lost all their moisture, and become like those vegetable 'crisps' that have become a fashionable alternative to the potato ones. It is very important that they are completely dehydrated before they are put into a Kilner jar, or something similar, for storage. If there is any damp they will go mouldy and ruin the rest of the crop. If they are completely dry, I push down hard and cram a dozen big ceps into quite a small jar. A handful removed and reconstituted a month or so later will lift a mundane dish into a gourmet treat.

I do not know why many people are suspicious of fungi. Could it be because they 'just appear from nowhere'? Is it because a few of them are so poisonous? Maybe it is because they have an association with rot and decay – that green bread forgotten in the corner, a bad apple dotted with grey, dusty spots. Yet without fungal associations plants could not grow healthily; without saprobes cellulose and lignin would swamp the world; without fungal antibiotics gangrene would have been as terrifying now as it was to the Knollys. Yet still I find a trampled mass of orange-brown glistening inkcaps (*Coprinopsis micaceus*) in the wood, squashed deliberately by a passer-by, as if in revenge for some crime. I can only think of the crime of being strangely beautiful.

Elm story

One of our cherry trees is sickly, its crown thinner than the others, and its leaves have tumbled off prematurely. Nearby, I discover a log from a tree that has fallen some while ago, and under its peeling bark are stout black threads, resembling thin electric cables or bootlaces. I know them well. Black rhizomorphs of honey fungus (*Armillaria mellea*)[8] are a sign that a tree is doomed. They extend from one tree to another, exporting death to the neighbours, and my cherry is the next victim. Once the roots have been killed, the honey fungus lives happily on the dead tree, eventually producing dense clusters of pleas-ant-looking yellow mushrooms as autumn progresses. That sick cherry will have to come down. Honey fungus has left behind the benign habits of its fellows to become a parasite. Instead of breaking down dead tissue it attacks living plants. Perhaps such malign fungi provide the real reason some people find them scary; and it is not an uncommon habit. Its extreme expression is when one fungus para-sitises another fungus. By the path I find an elegant little bonnet (*Mycena*) that has turned into a fuzz-ball on a stick. The original small pointy cap is covered with a ball of stiff threads sticking out like an Afro hairdo: a pin mould, *Spinellus fusiger*.

Further along the same path, fungal damage gets more serious. Several dead tree trunks are slowly decaying; one or two still stand, but others lie awkwardly on holly bushes they flattened as they fell. These wych elms (*Ulmus glabra*) have died as a result of another pathological fungus, Dutch elm disease (*Ophiostoma novo-ulmi*). The name is not a slur on the people of Holland: it acknowledges that Dutch scientists identified the organism responsible for the disease in 1921, since when it has devastated elms all over Europe. The fungus is spread from tree to tree by a small bark beetle (*Scolytus*) which is not interested in saplings; only more mature trees are vulnerable. Once infected with the fungus the whole water-conducting system of the tree clogs up with fungal mycelium, and the inevitable death of the crown is horrible to behold. Trees are reduced to bleached skele-tons. *The New Sylva* tells us that twenty-five million trees died in Britain after 1967, and thereby diminished irrevocably the rural land-scapes once painted by John Constable.

The loss of the material for the seats of Windsor chairs might seem a small thing in comparison, and it is some time since elm has been employed in the manufacture of water wheels and buckets, but we can grieve for a thousand panoramas that will never return. I do not have any evidence for English elm (*Ulmus minor/procera*) in Lambridge Wood, as it is a species that prefers hedgerows and copses to thick woods, as John Evelyn noticed in the original *Sylva*. However, in our wood the wych[9] elms are not all dead. One rather substantial tall tree survives, and there are a good number of small trees around the Dell in the middle of the wood. They are subtle trees that do not blatantly announce themselves; it was a while before I even noticed them. Their elongate, oval-shaped but pointed leaves are coarsely jagged-edged, and have a curiously lopsided base, resembling two ears slightly out of kilter. I have noticed that elm leaves differ from those of every other tree in the wood: they have variable sizes. A beech leaf is always a beech leaf and always about the same size, whereas wych elm leaves from a young tree in understory are much smaller than the yellowing examples that are now fluttering down from the canopy of my biggest tree. These must be four times as large – the size of lime leaves. It is as if wych elms are able to trim their photosynthetic spread to match the light available.

Wych elm is more resistant than English elm to Dutch elm disease. English elm reproduces almost entirely by suckers; its 'offspring' are genetically identical to the parent. Indeed, there is little genetic variation in the whole British population, which makes the species particularly vulnerable to disease. Wych elm reproduces by seed, which allows for enough cross-breeding for natural genetic variation to build up resistance to the fungus. Elm flowers are tiny, little more than pink stamens, but they make a brief show in early spring before the leaves have unfurled, as if the elm branches were putting out tiny streamers. Bunches of small papery wings clothe the twigs later, producing fussy ruffs behind the leaves; each wing encloses a single small seed. Even when a larger tree succumbs to Dutch elm disease – as several of mine must have done – new trees can readily regenerate from seeds. I fervently hope that my big tree is one of those that have cheated the fungus from its damaging livelihood, and that my many young trees are the first of a new, stronger generation.

Mind you, even dead elms can bring forth bounty: an extraordinary pink mushroom – the wrinkled peach, *Rhodotus palmatus* – is a prize that has arisen from my dead elm trunks; and it is a harmless saprobe. Nothing in nature is without compensation. But are any of my trees safe? Even the ash trees are threatened by 'die back', another fungal disease (*Hymenoscyphus pseudoalbidus*), that was only identified in 2006.[10] I have been nervously inspecting the ash trees in Grim's Dyke Wood for signs of this new pathogen, and there are none that I can identify. But a certain dread is abroad. Could this uneasiness lie behind the trampling of innocent mushrooms by aggressive boots?

Bee and spider

Lawrence Bee has made his last visit to Grim's Dyke Wood in search of the spiders he loves. He must be inured to jests about his surname slightly missing the entomological mark. Spider-collecting is like fly-collecting in reverse. Instead of chasing after the quarry with nets, spiders living in trees and shrubs are encouraged to drop out of their hiding places by beating the branches and collecting the spiders as they fall into an inverted contraption, a kind of mesh umbrella. It makes for a curious spectacle, as if invisible fruit were being harvested from inappropriate trees. Spiders are arachnids, not insects, so they cannot fly away, although some of them run quite fast. Different kinds of spiders lurk in the darkness of the woodpile or under logs. Others again spin their webs in the bramble patches. All of them are predators. They have been playing the arthropod version of cat-and-mouse with insects for four hundred million years, and evolution has shaped them with exquisite precision.

Suspended between the blackberry vines are archetypal spiders' webs, each orb as regular as if dictated by a geometer, yet every one as individual as if spun by an artist. Sitting motionless under the web is the common garden spider, *Aranaeus diadematus*, now at her maximum size and full of eggs. It is easy to spot a white cross on her abdomen that identifies this species. She has woven her web during the night. Low autumn light catches microscopic beads on the web, the stickiness that traps any unwary flying insect. All bundled up in silken thread in the corner of the web is a previous catch – maybe even a bee. The spider injected venom into the insect that served to pre-digest her meal as well as kill it. Earlier in the year I had watched the nervous courtship of a smaller, slimmer male spider strumming the web to pacify the object of his desire. Males have to absorb sperm on to their palps (special appendages) and then insert the package into the females to ensure fertilisation of the eggs. They may get eaten for their trouble. There is no mercy in the arachnid scheme of things: protein never goes to waste. In May, I had played with the tiny spiderlings that had hatched out from an overwintering egg package. They hung together in a golden ball until touched, when they splayed out in all directions along nearly invisible threads, like an exploding star. Only one or two of these tiny creatures would make it through to become a succulent web-builder like the one before me now.

More than thirty species of spiders have been found in Grim's Dyke Wood, providing a whole catalogue of entrapment and wiliness. A variety of sheet-weaving spiders make white webs looking like untidy hammocks; they can be found decked around herbs or in trees. They lack the formal beauty of the orb webs,[11] but they are as effective in catching smaller insects. When dew condenses on them they suddenly become easily visible. On sunny days they can apparently vanish. Tangle-web spiders have three-dimensional webs, finer than thistle-down, that seem to lack any logic at all, as if they were stranded wisps of candyfloss; but they are snares of subtle complexity. One of these species in the wood is the mother-care spider (*Phylloneta sisyphia*). She guards her spherical egg sac, feeds the hatchling spiders on what she regurgitates, and later on what she catches, and finally dies so that her offspring can make a meal of her dead body. The funnel weaver (*Tegenaria silvestris*) makes a trap around the hole in which it lives composed of threads that are not sticky, but are teased to snag the legs of any passing item of prey – which is then dragged into the lair to be consumed. Sac spiders use their silk to make a hideout for themselves inside rolled-up leaves or under bark which they occupy during the day, only to emerge at night to stalk their prey directly. With eight specialised eyes to scope the world they do not need the ruse of a sticky web to secure their food. In the wood there are three species of *Clubiona* with this mode of life, each with slightly different habitats on trees or on the ground. An uncommon little crab spider (*Diaea dorsata*) lurks on holly leaves to grab its prey: the front parts of the animal are all coloured green, and its front legs are long and held out like a defensive shore crab. Maybe it would be a quicker death for an insect to fall into the strong jaws of the four-eyed jumping spider, *Ballus chalybeius*, which can both stalk and pounce. To complete this catalogue of carnage, let us bring out a pirate spider (*Ero furcata*) that pursues and eats the orb-weavers. Nobody of bug size is safe in Grim's Dyke Wood.

On one of my regular ambles through the wood I spot a curious object lying among the litter. A piece of wood – probably beech – but inflated like a bladder, and approximately the size of a tennis ball. I imagine it must be some kind of gall produced on a tree branch in response to an attack by an insect, or maybe an infection. In any case it is now hard and hollow, like the flint that held the Iron Age hoard. This flask holds a spider. A perfectly round hole at one end is lined with silk. This is where the lace-webbed spider (*Amaurobius fenestralis*) hid from its enemies during the day before coming out to hunt at night. Such a secret passage hidden in wood must be added to the collection: I have no other example of a portable lair.

An orb-web spider (*Aranaeus*) spins its subtle trap in the wood.

Many spiders will die with the first frosts. Near the end of October the autumn colours are well advanced. Seen from the Fair Mile the wood looks all gold and russet. Only the cherry leaves provide deep-red tones; the English maple and the wych elm leaves are lemon yellow, and English oak a dull brown in its annual decline. Golden, orange and brown beech dominates. Even without scarlet, the Chiltern woods have provided the decoration for a thousand chocolate boxes; but though the sight is thrilling, it is perhaps just too obvious to have inspired great art. Constable, Paul Nash, even David Hockney have avoided the subject. Inside the wood everything still looks quite green. The leaves have changed colour at the tree crowns first, while those on branches nearer the ground are still verdant and remain attached. The sun in the canopy provides a golden painted roof reflecting down into the gloomier woods below. A few leaves flutter down from the high branches. Husks and nuts still lie on the ground to crunch underfoot, reminding me of the bombardment I had received earlier in the month. Now these falling leaves would hardly disturb a spider on its web.

8

November

Little shots and pheasants

Frost and wind have detached nearly all the leaves from the beech trees. Only young saplings hang on to theirs, as they will all winter. The floor of the wood is now all orange-brown, and the tree trunks parade upright in their stripped ranks. The light has changed drastically in the last week; hard and clear, today it accentuates openness and airiness. I realise that I had missed seeing the sky; there was something oppressive about the dense canopy that shut out so much of the world. Now the wood has been readmitted to the wider countryside. To the south I appreciate the gentle rise and fall of distant hills – other woods with their own histories, arable land sown with winter wheat, a patchwork of usage that has been the pattern in the Chiltern Hills since Saxon times. All the beech trunks shine as if they have been newly polished. A small biplane from White Waltham airfield is retreating northwards overhead in the direction of Fawley Court. I wonder why the growl of a turbo-prop engine always sounds so melancholy; more a sad, modulated groan as if in mourning than any kind of machine.

A pair of red kites circle high overhead, the distinctive profile of their forked tails highlighted against a porcelain sky. They cry to one another with a shrill urgency: wee-wee-woo, wee-wee. The two wheeling birds are searching for carrion, and they are buoyed effortlessly on thermals. Kites only came back to their home in the Chilterns twenty-five years ago, after their extermination in the nineteenth century; it was a

deliberate reintroduction. I recall my amazement when I first saw one
– it seemed then an impossibly large and exotic raptor, as if it had flown
in from the Caucasus or beyond. Now these huge birds are everywhere.
Kites prefer to keep out of the wood; open skies are their domain. Their
cries are heard often when they cannot be seen.

Lambridge Wood is home to a pair of common buzzards. This
species too had been eliminated, but came back of its own accord,
following the kites. Buzzards are burly, serious birds, but they display
surprising agility when they fly through the densest parts of the wood.
I watch a stately bird sitting motionless on a broken branch command-
ing a view of the woodland floor. Its eyes glitter. Any unwary wood
mouse would quickly provide it with a useful snack. The buzzard's
only rival hunter inside the wood is a tawny owl, which I have heard
spookily woo-wooing but have never seen. Since it is largely a nocturnal bird I imagine that it does not cross paths with buzzards on many
occasions, but I do know this stealthy hunter has been through Grim's
Dyke Wood. I discovered a dark, bean-sized owl pellet composed of
its regurgitated, undigested waste. When I dissected the object under
a binocular microscope it pulled apart easily. I found matted hairs,
vertebrae and a few tiny teeth that were all that remained of a mouse
or a vole; glassy claws of some small bird.

A fine cock pheasant is standing on top of the woodpile. He is by
far the showiest bird ever to grace the wood: white collar like that
worn by any country vicar contradicted by a red head and breast and
blue neck, recalling some grandee prelate's finery. His fat body is
warmly speckled and barred brown as if it had been generated from
the beech litter by transmutation. This posing popinjay reminds me
why kites and buzzards were formerly so persecuted. Beech woods on
estates were, and indeed still are, assets for shooting, and all birds of
prey are anathema to gamekeepers. Lambridge Wood has no regular
shoot in the twenty-first century, but if I listen hard I can just make
out the sharp cracks of rifles coming from over Fawley way, or maybe
as far as Stonor. Pairs of birds are already hanging in the window of
Gabriel Machin, butcher to the discerning of Henley. The cock pheasant in our wood is an escapee, though he does not know it.

I reflect not a little ruefully that pheasant-shooting is overall a good thing. Were it not for this sport of the well-to-do or well-connected, many more beech woods would probably have been cleared by now and put down to barley. When timber lost value, status sports did not. In 1854 George Jerome of Henley 'was charged with trespassing in Lambridge Wood on 24th Dec., in search of game'.[1] Poor man, he was probably trying to get something to put on the Christmas table. Lord Camoys of Stonor fined him five shillings (plus ten shillings costs), or twenty-one days' hard labour. George Jerome probably had no choice but to accept the latter.

Big shots and Feoffees

In early 1643 the pigeons in Lambridge Wood suddenly broke into panicky flight. A cannon was discharged in Duke Street in Henley, killing four Royalist soldiers.[2] The 'Henley Skirmish' marked a violent climax to the period when the town became a battlefield in the Civil War. In the tussle between King Charles I and Parliament, Henley was in a vulnerable position for the same reasons it had become such a successful commercial centre. The town lay between Oxford, which the King had established as Royalist capital in 1642, and London, which was staunchly with Parliament. Henley was an important port and river crossing – its bridge was destroyed and rebuilt several times during the years of turmoil. After the Battle of Edge Hill on 23 October 1642, the town was battered in a bloody tug-of-war between the opposing sides. Reading was initially held for the King, and Prince Rupert – Charles's ablest commander – occupied Henley, while a large troop of his horse were quartered at Fawley Court, less than a mile downstream towards Marlow. The Fawley estate is neighbour to Greys – and our wood – on the north-eastern side of the Fair Mile. Seen on a clear November day that part of its grounds called Henley Park rises clearly to a similar height to Lambridge Wood, though only the upper part of the flanks of the hill opposite us is wooded today. It too became part of our history.

At the time of the Civil War the Fawley estate belonged to the splendidly named and commensurately distinguished lawyer Sir Bulstrode Whitelocke. He was a Deputy Lieutenant of Oxfordshire, and a Parliamentarian, though a moderate and sensible man who was reluctant to see any harm happen to the King. He had removed himself to London when Henley was taken over by Prince Rupert and his henchmen. A thousand men of the King's horse trashed his grand house and ruined his estate. 'Divers writings of consequence and books which were left in my study, some of them they tore to pieces, others they burnt to light their tobacco ... they broke down my park pales, killed most of my deer ... and let out all the rest, only a tame young stag they carried way and presented to Prince Rupert ... They ate and drank all that the house could afford ... they likewise carried away my coach and four good horses, and all my saddle horses, and did all the mischief and spoil that malice and enmity could provoke barbarous mercenaries to commit.'[3] The cries of these rowdies would have carried to our wood from the other side of the valley, and escaping deer would likely have fled towards Greys Court. It is even conceivable that unbridled looters purloined feed for their animals from the neighbouring estate, for the Knollys were well known to be devoutly Protestant. Well-armed Royalists had also garrisoned the next big property upstream, Greenlands House, half a mile onwards from Fawley in the Marlow direction.

Some relief was at hand. A force of Parliamentarian horse and foot under the command of the Earl of Essex occupied Henley on 23 January 1643, and the 'Henley Skirmish' happened later that night. The Royalists retreated back towards Reading. The whole action eventually cost a dozen lives. With Roundhead troops billeted on the town, and Essex occupying Phyllis Court on its northern edge, the townspeople were likely as harried now as they had been when the King's men were in command; demands for provisions and levies to further the campaign were unrelenting. St Mary's church by the bridge, for four hundred years a place of worship, was demoted to become a stable for horses. Earthen fortifications were thrown up about the town. The Roundhead soldiers who now occupied what was left of

Fawley Court were no better behaved than their Cavalier predecessors, and Bulstrode Whitelocke complained indignantly of the damage they did to his woods.

The river trade, Henley's lifeblood, became perforce sporadic at best, not least because the King's troops occupied redoubts both upstream and down. It was a time of violation and privation in equal measure. The death rate in Oxfordshire as a whole more than doubled as disease and malnutrition took a further toll. Finally, after several attacks and the expected retaliations, the Royalist forces downriver at Greenlands House surrendered in June 1644, on condition of safe passage to Nettlebed with their arms and horses. They must have retreated past our wood, passing along the Fair Mile and onwards up the old route towards Bix at the top of the hill. A poacher taking advantage of the disordered times and lying low in Grim's Dyke Wood would have heard the clatter of horses' hooves and the coarse cries of the carters encouraging their charges up the steep and rutted road that broached the Chiltern Hills above Assendon, a thoroughfare so ancient that it had been carved into a deep holloway. The old route to Oxford carried a melancholy procession of dispirited Cavaliers; and still the Civil War was far from over.

If that same poacher had been in the wood on 27 April 1646, he would have witnessed an extraordinary parade passing the other way. King Charles I had escaped from Oxford in the guise of servant to two friends who were now his travelling companions. The trio crossed the Chiltern Hills from Nettlebed, en route to Henley and Maidenhead. Their future plans were still nebulous. The King had sacrificed his lovelock – symbol of the Cavalier – and was dressed in a Montero[4] hat. With the help of a false Parliamentarian pass, the disguise worked well enough to get the party past the roadblock into Henley; tipping the guards twelve pence may have helped. Nonetheless, the experience must have been humiliating for one as vainglorious as Charles. His admirer John Cleveland may have been overstating the case when in his poem 'The King's Disguise' he described his hero as 'The Princely Eagle shrunk into a Bat', or even 'A pearl within a rugged Oyster shell'.

A seventeenth-century soldier sporting a Montero.

Nonetheless, the experience of being 'so coffin'd in this vile disguise' symbolically marked another stage in the King's ousting that culminated in his execution by the axeman on 30 January 1649. As early as the summer of 1646 Sir Bulstrode Whitelocke, by then Governor of Henley, had mobilised local people and troops to dismantle the military works in town and reopen the bridge across the Thames. It must have been a profound relief to merchants and artisans alike to see the end of quarterly assessments for taxes, and arbitrary requisitions of this and that, not to mention the removal of the ruffians among the soldiery. During the Commonwealth and the Restoration, normal trading life could resume.

Our wood had borne silent witness to history being made in the valley below it, and now returned to its traditional employment. The value of beech wood for fuel continued to rise. The diarist Samuel Pepys referred specifically to the trade in Henley around 1688, noting that 'beech woode, [which] is said to burn sooner, clearer, freer from sparkle, and to make a better coale, yt will keep fire longer than those of oake, though oake last longer in ye burning than beach [sic], the measure and price being … ye same or near it.'[5] As Secretary to the Navy Board, Pepys was intensely aware of the importance of timber in all its manifestations. Dr Robert Plot had observed that 'In the Chiltern Country they fell their Under-wood Copices commonly at eight or nine Year's Growth, but their Tall-wood or Copices of which they make tall Shids, Billet, Etc, at no certain time; nor fell they these Woods all together but draw them out as they call it, almost every year.' Coppiced beech had to be cut to the right lengths for the London market; the brush bound into faggots, perhaps to supply the brick-works at Nettlebed or Reading; hazel poles harvested for local use in building or fencing. There must have been discussion about whether to take down one of the big trees soon, or wait for the market to improve. Even the return of plague to the capital in 1665 hardly inter-rupted this routine, though the Henley bargemen were forbidden to import goods from London for a while. The cycles of woodland regeneration went in harmony with an ebb and flow of wildlife, the clearings full of summer butterflies, and regular felling encouraging a wealth of wildflowers; only a remnant of this biological richness still prospers.

There was real money to be made from woodland in the seven-teenth and eighteenth centuries. When Sir Bulstrode Whitelocke consolidated his estate in 1637–38 he bought what is now Henley Park, to the east of our wood on the other side of the Fair Mile, which at the time comprised mostly woods and coppices, and continued to be managed in the traditional way. In the 1630s Sir Bulstrode sold ten thousand loads of firewood from his total estate to a woodmonger in London, netting him a profit of £3,000, a huge sum by any measure. In 1672 Bulstrode settled Henley Park on his son William, who then proceeded to convert much of the woodland to arable land. He leased the park for £2,000 to John Cawley, rector of Henley, and to John Taylor, with permission to clear the woods; a hundred acres was grubbed up and sold on for a handsome profit. By the second decade of the eighteenth century Henley Park extended to about four hundred acres of enclosed land, comprising patches of woodland remaining on the steeper parts, and elsewhere open ground that had been cleared, presenting a landscape generally like the one we still see from beyond our wood.

A large bite was taken out of Lambridge Wood at a similar time, and it came within a whisker of our piece of it. Just to the far side of the barn and cottage at the northern end of Grim's Dyke Wood a thin strip of several fields extends towards Henley on the flanks of the hill. This forty-four-acre patch was once woodland just like our own. In 1658 William Knollys of Greys Court granted a ninety-nine-year lease on the wood as it then was to Thomas Goodinge of Henley, gentleman, for £22. In 1681 William Hopkins left £300 in his will to be invested in land as charity to the poor of St Mary Magdalene, the parish church of Oxford, 'to be laid out in bread, and given every Saturday at evening prayer'.[6] The piece of woodland near Henley became the basis of the Lambridge Charity. The proper administration of land charities requires the oversight of many Feoffees (trustees), and this generates a paper trail from the first, in contrast to our own piece of woodland, where plausible inference is all I have. One of the more legible indentures of 1797 tells us that the Feoffees, all Freemen of Oxford, included John Parsons, mercer; Thomas Pasco, druggist; Thomas Wyatt, baker; William Hayes, bookbinder; James Costar, tailor; Edward Rusbridge, cordwainer; Thomas Looker, grocer; and William Winter, upholsterer – thus providing a neat encapsulation of the state of business in the county town.

The forty-four acres yielded income for the needy for nearly two hundred years, and the Oxford Record Office, housed in a deconsecrated church on the Cowley Road, has a great sheaf of papers recording changes of lease and the occasional bankruptcy. Among them is a somewhat scrappy fragment dated March 1707 that tells us of 'An agreement between Rbt Waters of Henley & Philip Seale & William Brookes of London for the Roots and Runts grubbed & to be grubbed on the estate called Lambridge.' The roots, and indeed the runts, were converted to charcoal, and £3 was paid for the privilege of doing so. The felling and sale of the beech wood itself, much of which was the business of the Waters family, was followed by thorough clearance,

Ancient Chiltern countryside: the view westwards across the Assendon Valley from Henley Park with our wood and the site of 'the murder cottage' on the skyline.

The elaborate monument to the Knollys of Greys Court in Rotherfield Greys parish church. The Knollys were a powerful family in the Tudor court.

The manor house of Greys Court, Oxfordshire, today, which is much as it was at the time the Knollys managed our wood.

The Stapleton family. In the foreground, the unmarried Stapleton sisters who occupied Greys Court for much of the nineteenth century. Painted by Thomas Beach, 1789 (Holburne Museum, Bath).

The pelargonium variety 'Miss Stapleton', named for the senior Stapleton sister, who was an enthusiast for these plants. Her greenhouse still remains at Greys Court.

Document recording the granting of rights to grub up the 'Roots & Runts' of beech trees cleared just below our wood, in aid of the Lambridge Charity (Oxfordshire Records Office).

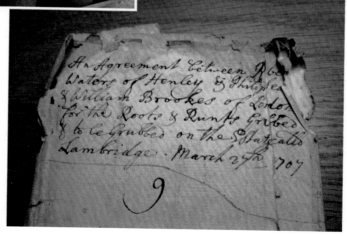

The black-and-white livery of the magpie toadstool is distinctive among the ink caps, and it is a typical species of beech woodland.

The stinkhorn fungus (*Phallus impudicus*) lives up to its name. The flies are enjoying a malodorous 'soup' which helps spread the spores.

The sulphur polypore grows out as soft brackets from dead cherry wood. It becomes more leathery as it matures.

Looking upwards towards the leafy canopy through a crowd of glistening ink caps growing on a rotten beech stump.

Rhodotus palmatus is an uncommon and beautiful mushroom with a wrinkled cap skin. It grows on fallen wych elm trunks decaying in the wood.

The red slug *Arion rufus* on its way up a tree in the wood.

A glossy black dor beetle, *Anoplotropes*, in search of herbivore dung in which to lay its eggs.

An uncommon beetle, *Oedemera femoralis*, with unusually long wing cases (elytra).

A strikingly coloured longhorn beetle, *Rutpela maculata*, showing the long antennae. Their larvae feed on wood.

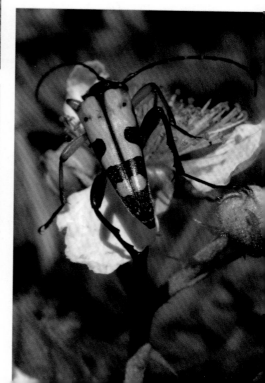

A sexton beetle, *Nicrophorus humator*, coleopteran undertaker. This specimen is carrying a tiny mite which will hitch a ride to a promising carcass; the beetle is unaware of the passenger on its back.

Alistair Phillips at work at his lathe turning a wild cherry-wood bowl from timber derived from one of our felled trees.

Small-scale charcoal-burning in progress in an old oil drum. Charcoal was one of the important products of the wood in medieval times.

A flint derived from the chalk at the Fair Mile, with a fanciful resemblance to a sitting cow (the 'rother' of Rotherfield) – one for the collection. Pure flint like this makes good stone tools and the best glass.

The author making notes – sitting on a fallen cherry branch.

Jan Siberechts' 1698 painting of Henley
showing flashlocks in the foreground,
while timber that could have come from our
wood is piled up by wharves near the town.

An autumn leaf of wych elm resting on an
elm trunk shows its typically asymmetrical
leaf base.

In early spring, male wych elm flowers are little
more than bunches of colourful stamens.

The drawers of Philip Koomen's cabinet hold the treasures of the collection. The dormouse nest is backed by my leather-bound notebook, and our own green glass.

Lichens decorate fallen branches in the winter. This one, *Parmelia* (shield lichen), can tolerate pollution. Others are very choosy.

The Koomen collection cabinet back in the wood, with the cherry planks that gave it birth stacked up behind it.

Mosses form delightful cushions in wetter, colder times, like this bank haircap (*Polytrichastrum*).

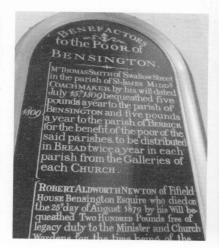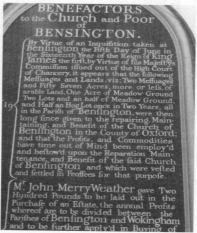

Charity boards in St Helen's church, Benson.

and only then was the land fit for farming. By 1770, a ten-year lease granted by the Feoffees to James Brooks of Henley, gent., of 'three closes of arable land called Lambridge', proves that clearance had proceeded further, but up to the nineteenth century the phrase 'which were once woods' is still appended on documents to the description of the forty-four acres of land, even after a farmhouse was built near the top of the ridge just below the trees, well to the east of our piece of Lambridge Wood.

The memories of woodland past evidently linger on. Lambridge Hill, as it is called on the Ordnance Survey map, still looks like a piece of land snipped out of a patch of green forest. The charity was wound up in 1882 with the sale of the farm for £14,000 to Colonel W.D. MacKenzie, then owner of the Fawley estate, thus bringing that estate still closer to Greys Court. Today, the much-elaborated farmhouse is occupied by a reclusive envoy from somewhere in the Middle East who has fenced off the former woods and later arable farm as a private deer park. I reflect that our own piece of woodland again had a narrow escape. If the eighteenth-century clearance had proceeded just a smidgen further, our precious link with deep time would have been irrevocably severed.

After the Great Fire of London in 1666, Britain's greatest Renaissance man, Sir Christopher Wren, set about designing the rebuilt St Paul's Cathedral. Work on the architect's masterpiece began in 1675; Wren's son laid the last stone at the top of the lantern in 1710. One of the consequences of the great conflagration was a new appreciation of the importance of building fireplaces with efficient flues. The demand for wood as a domestic fuel continued, but coal would begin to be regarded as a more efficient source of heating, leading eventually to a change in demand for one of the routine products of Lambridge Wood. The easterly part of the wood was then part of the Badgemore estate (later the home of the banking Grotes), which leads to another local connection with the Great Fire. Wren's master carpenter was Richard Jennings, and the same year that the cathedral was 'topped off' Jennings bought and remodelled Badgemore House and, according to Emily Climenson's 1896 *Guide to Henley*, had the bricks and scaffolding remaining from St Paul's brought upriver to do so. A little piece of one of England's greatest buildings found its way to our neighbourhood. A remarkable scale model of the west front of St Paul's belonging to Jennings found its way into Shiplake church – even though Jennings himself was buried in St Mary's church in Henley. It is now back in the cathedral. In 1712 Jennings and Wren were close allies defending a lawsuit brought against them by the Dean of St Paul's alleging financial mismanagement, from which they emerged largely exonerated.

I too share an esoteric connection with Sir Christopher Wren. He was one of the Founding Fellows of the Royal Society, the oldest academy of science in the world, which was formally established with a lecture that Wren gave in 1660. On election, a new Fellow has publicly to sign a big book in Indian ink. Wren's signature is on page one. I signed the very same book when I was elected a Fellow in 1997; out of sheer nervousness I made an inelegant blob. I have a bond with Richard Jennings through a signature that he must have known better than I.

Rot and renewal

It is still quite mild for November, but several days of cool rain have left a legacy of general sogginess. The old rotting log that sprouted sulphur polypore, or 'chicken-of-the-woods', back in springtime is looking ever more decrepit. It has fallen into two chunks, and any shreds of bark it once retained have disappeared. The wood beneath has not decayed evenly, but has broken down into a series of grey ribs made of harder heartwood, like a scrawny neck; some of the wood elsewhere has become almost crumbly, already on its way back to soil. I push my finger into it to test it. Then I take one half of the log and roll it over and away a short distance; beneath it is a hidden world. At once there is intense activity among many small creatures – they are scurrying away from the light as quickly as they can. They are suddenly visible to potential predators, and their instincts to flee the light are instantly on alert. Woodlice are the dawdlers among them; they amble in a mechanical fashion, like so many tiny wind-up toys, as they hide in crannies on the ground where the log had rested – and the soil is indeed a mass of holes and runways resembling a hidden labyrinth. A couple of bright brown centipedes are off quick as a flash to hide under stray beech leaves, all scuttling limbs and alert antennae. A shining black ground beetle takes off with equal alacrity. An earthworm (*Lumbricus*) whips itself into feeble coils. I have disturbed this unlit habitat with my sudden intrusion, exposed a little world secreted within the larger world of the wood, itself just one of many ecological realms within one small island. I need to get closer to this living dungeon with its dark secrets.

On the underside of the log, now lying exposed to the world, are two whitish irregular patches smaller than my hand. They resemble parchment pressed on to the surface of the rotting wood. Both are fungal fruit bodies of amorphous form: they digest wood. Such fungi lurk on the underside of damp wood where humidity is high, feeding off the materials that make up the cell walls of the wood. They are the unappreciated heroes of decay, planing off the lower surfaces of logs until they sink inexorably year on year towards the ground. My lens reveals that one them is actually composed of a complex maze of minute creamy walls. *Schizopora paradoxa* is one of the commonest of these 'resupinate' fungi, and secretes enzymes that allow it to digest tough lignin – only fungi can do this vital trick.[7] When it has finished its work on small branches elsewhere in the wood they weigh almost nothing; they have become ghosts of their former selves, and like all self-respecting ghosts they are white (at least, the wood is white), which is the tint of the cellulose left behind. The other patch is a much smoother species of *Hyphoderma* performing a similar task. When I examine these patches under my microscope back home I soon see that much of their surface area is devoted to shedding spores, and from the way the spores are borne (on four-pronged cells called basidia) I recognise that these species are, despite appearances, related to conventional mushrooms. They are fungi that have given up look-ing like mushrooms in favour of lying doggo in the dark: 'white rotters' to those in the trade. Although there are none of them on this particular log, 'brown rotters' digest cellulose and leave the lignin, so that the infected wood becomes red-brown, dehydrated and cubically cracked; several examples are found on my conifer pile. Between them, these inconspicuous fungi recycle just about all the wood in nature.

They are also at the base of a food chain. A strong magnifying glass reveals another world. The surface of the fungal sheet has what look like black, moving punctuation marks wiggling upon its surface. I recognise them as tiny mites (*Xenillus*), diminutive relatives of the spiders. There are several species of these minute creatures; some even carry spikes on their backs like fantastical monsters recruited from *Star Wars*. They are grazing on fungal material, including the spores. Equally small creatures are alongside the mites, looking much paler and elongate; they have six legs like any conventional insect, but are flightless. If disturbed, they are liable to 'ping' out of view because they have an escape mechanism that gives them their common name of 'springtail'. Science calls them Collembola, and how exactly they relate to the flying insects is still debated, although they have been regarded in the past as a primitive group that had not acquired the useful piece of aerial apparatus of their distant cousins. No matter: they are among the most abundant animals on earth – in damp vegetation there may be as many as 100,000 per square metre – and the world under the log provides them with all the detritus and bacteria they need.

These tiny animals are about as small as I can comfortably make out in the wood with my naked eye, and to see any details magnification is essential. Mites and springtails are food for other creatures, though probably not for a large, pulsating grub that looks like the larva of a fungus-eating beetle. I get a momentary *frisson* when another tiny animal runs in under my glass – it looks exactly like a reddish scorpion. As my sense of scale reasserts itself I remember that it must be only a few millimetres long, but just for a moment my heart went pit-a-pat. Now I notice that it doesn't have the sting in the tail either, though it does carry a fine pair of pincers at the front. I don't doubt that it would make short work of mites and springtails. This fearsome, if diminutive, predator is a pseudoscorpion, *Chthonius ischnocheles*, and demonstrates that the complexity of the name is inversely proportional to the size of the animal. It has been the subject of a scientific paper specifically based in Lambridge Wood,[8] where two other pseudoscorpions (*Neobisium muscorum* and *Roncus lubricus*) are recorded. *Chthonius* is derived from the Greek for 'subterranean', so it is an appropriate name, if hard to get one's mouth around. It inevitably recalls H.P. Lovecraft's sinister and deeply buried civilisation of Cthulhu.

And indeed we would find the dark world under the log quite alien, for the hunters there lock on to chemical smells detected through the slightest twitch of an antenna; it is a lightless realm of stealth and subterfuge. Here live centipedes, a group of fearsome predators: with specialised limbs around the mouth, they incapacitate their prey using their special venom claws before shredding their victim to pieces. The red-brown runners I notice scurrying for cover are classified in *Lithobius*, with three species in the wood, of which the largest, the banded centipede *Lithobius variegatus*, bearing striped legs, is a good indicator of ancient woodland in Oxfordshire. These legs stick out like oars from a Viking ship bent on pillage, but unlike any ship you have ever seen the body can flex. The back end of this animal is confusingly like the front, because the centipede's antennae are mimicked by a pair of 'caudal furcae' located close to its back end.

A longer, yellower, and altogether thinner centipede, much more generously endowed with up to eighty-five pairs of legs (*Stigmatogaster subterranea*), moves more slowly when disturbed under the log, and bends itself into elegant curves as it negotiates its escape, like a self-twisting rope. Slender, flexible and active *Geophilus* species possess limbs that tot up somewhere between the fifteen or so pairs of the brown centipedes (*Lithobius*) and the numerous ones of *Stigmatogaster*; and finally there is small and elegant *Cryptops hortensis*, a species related to giant and very poisonous centipedes from South America. What *none* of these centipedes has is the 'hundred legs' promised by their common name. They always have an odd number of pairs of legs, matching the number of rings into which their bodies are segmented. Each species must have a favourite prey and favourite schemes for hunting them, but we will not know the details until we can learn to see intimately into their secret, blacked-out world.

Still subtler ploys are played out where no eyes can follow. Fungal hyphae – the living threads of the organism – quest through dead beech wood in search of nourishment, but dead wood lacks nitrogen compounds that help the fungus grow. Within rotten wood also live many kinds of tiny nematode worms. I often come across them under the microscope thrashing about in their death throes – tiny, all-but-transparent tapering sausages – which simply cannot be identified without sophisticated modern molecular techniques.[9] These little worms do contain useful quantities of nitrogen. Fungi have evolved techniques for capturing and consuming unfortunate worms by spinning lasso traps with their hyphal threads: the traps tighten on the tiny worms as they pass. Sucking on the dead bodies, the fungus can now break down yet more wood with renewed vigour, bolstered with a nitrogenous 'fix': murder in the dark.

Woodlice eat wood. They are not insects, as a casual observer might assume. In fact they are crustaceans, second cousins to shrimps and crabs. They are also called slaters. Their close relatives (marine isopods) live in the sea, and some of the bigger ones look a little like my own favourite animals – the extinct trilobites – although the resemblance is superficial. Uniquely among crustaceans, woodlice have been abundantly successful in their colonisation of land. Nobody has much to say against them – they are not like bedbugs – but they are not particularly endearing either. No bestselling children's author has published 'Willie the Woodlouse' or 'Susie the Slater' books. Munching on wood is not a very glamorous option. There are five common species, and all of them are present in our wood, skulking under rotting logs, milling about on their fourteen pairs of legs. They do an extraordinarily important job in processing fallen organic material: their droppings make up an important part of 'frass', a mass of minute pellets that provide a kind of compost on which bacteria thrive, and are all part of returning hard, intractable wood to the soil. Under my lens I spot tiny white woodlice, and they are indeed their babies. Fertilised eggs are carried under the mother's body until they hatch into miniature woodlice all ready to go. Woodlice are nothing less than self-perpetuating reprocessing machines.

Millipedes are mostly longer, slender vegetarians, and might be mistaken for centipedes until you see that each walking segment of the long body carries not one, but two pairs of legs; and no, they do not have a thousand pairs of legs, any more than the centipede has a hundred – although the world record of 750 pairs on a Californian species was set as recently as 2012. Millipedes amble slowly along, in no hurry. The commonest of our species is a woodland inhabitant everywhere (*Cylindroiulus punctatus*) that looks like many another of those purposeful many-legged perambulators, and its common name – blunt-tailed snake millipede – does not help much to pin down its identity. Four other species are in Grim's Dyke Wood. One of them, the pill millipede (*Glomeris marginata*), rolls up for protection into a tight black shiny ball the size of a large pea, its segments interlocking until the head and tail come together. When I first saw one of these little spheres I wondered if it were some kind of strange seed, until it unrolled and walked away. Many of my trilobites could pull off the same trick, and probably for the same reason, providing another wonderful example of convergent evolution, like the ghost orchid and the Dutchman's pipe (pp.70–2). Another millipede, *Chordeuma proximum*, was something of a 'find', a westerly species making a rare appearance in the eastern part of England. Flat-backed millipedes (*Polydesmus*) look as if they were assembled from some kind of kit that clicks together to make miniature armoured trains trundling along at a regular pace. Dashing black carabid ground beetles, or the fat, three-legged toad that lives in the woodpile, might find them to their taste.

A list of animals and fungi could become tiresome, but is necessary to grasp the true richness of nature. Think of it as not so much an inventory as a catalogue leading to compelling and interlocking stories. The world beneath the rotting log is a small one, but it is marvellously complete. The cascade of life there comes ultimately from the sun. The photosynthetic work of a tree eats up the energy from sunshine for many years; as soon as the tree falls to the ground the construction begins to unwind. Fungi play the vital role. A beech log left behind from cousin John's felling is already dotted with hard brown spheres of *Hypoxylon fragiforme* ('beech woodwart'), a pioneer species that will be followed by a succession of others. I shall be looking for oyster mushrooms there next year. Beneath the log in the damp, dark places, the recyclers and degraders get to work. Different fungi from those on the exposed wood form their subtle patches, and their mycelium buckles down to unlock the energy stored in the wood. Worms tug organic matter down into the soil. Wood-eaters, and grazers of fungal patches, and then their predators, set up a food chain that is a lightless version, a dark parody, of the grass–herbivore–carnivore system that thrives in light and rain. Rot is creation in the underworld. That list, that catalogue,[10] is the *dramatis personae* of a kind of soap opera of slow decomposition, where sex and death, voracity and subterfuge, play out their measured parts in the life habits of dozens of species 'hidden away privily'.

I notice a pile of nibbled fruit stones ('pits'), no bigger than small peas. They are bleached white; at least twenty of them in what must have been a secret hideaway before I moved the log. For a few seconds I mistook them for rabbit pellets. They are surely from the wild cherry trees that grow nearby, and every one has a neat hole gnawed into it. I gather half a dozen of them to add to the collection. The holes are not more than a few millimetres across, with regular sloping bevels. They are the work of wood mice (*Apodemus sylvaticus*) seeking the nutritious kernels hiding in their woody cases. The little midden lies near a burrow in the soil that must have been roofed by part of the rotting log – this is Mouseville, Oxfordshire. The burrow continues under the adjacent log; it may well lead to a leaf-lined nest. I had seen wood mice a few times during daylight hours rustling dry leaves as they scuttled through the litter, all ears and cautious eagerness; pretty little grey-brown popeyed creatures. They really belong to the night, however, when they are out scavenging fat centipedes and bringing home beechnuts to store against hard times. The log seems to yield up more and more secrets; but it is time to move it back to its original position, and leave its underground inhabitants to their own multifarious devices.

Other mammals: I have only once seen in the wood that most voracious insectivore, and Britain's smallest and most primitive mammal, the pygmy shrew (*Sorex minutus*). It appeared in the leaf litter under the King Tree in August, an animated black ball of energy that soon shot off into hiding when it realised I was there. The little shrew is famous for having to consume more than its body weight of live food each day, and won't turn its pointy nose up at woodlice for supper, which I imagine as kind-of unpalatable crunchy biscuits. We bagged a field vole (*Microtus agrestis*) in one of our summer surveys. This is a very inquisitive animal and it is easily tempted into a mammal trap. It is short-eared and snub-nosed compared with the wood mouse, and its tiny beady eyes contemplated us with apparent unconcern before its release back into the wild.

I was surprised to find plenty of evidence for another creature of the underground, Mr Mole (*Talpa europaea*), an animal that I am unable to think about in an objective way since first seeing E.H. Shepard's illustrations for *The Wind in the Willows*. I did not expect that molehills could be thrown up from ground so densely packed with flints as that underlying Grim's Dyke Wood, but there are several places where the rewards in worms must have repaid the effort of digging. Large molehills around the big clearing are little more than heaps of stones. I imagine Moley sitting in an armchair underground, reading the newspaper and eating marmalade sandwiches. I have never seen his enemies the weasels and the stoats, though I do not doubt that they have passed furtively through the wood in search of wood mice and voles. After all, Kenneth Grahame called these predators 'the Wild Wooders', and Lambridge Wood is an approximation to the wildest, or at least oldest, we have left in England.

In November, late in the month on one of my regular, mooching visits, the wood was visited by a numinous presence. A brown hare (*Lepus europaeus*) loped through the trees, its implausibly long, brown-tipped ears twitching as they listened for any suspicious noise. The hare moved slowly, like a racing greyhound forced to mince around a suburban park. I had thought the hare a creature of wide fields, but here it was in the wood, questing for something – I still do not know what. No common-or-garden rabbit has ever crossed into our territory. The visit of this elegant animal, all legs and ears, made me feel curiously blessed.

More news from the underground

In the autumn of 1683 the wood pigeons were again startled into panicky flight. A witness reported that 'there was suddenly heard a strange and (for many years before not known) a most furious Commotion of the Air attended with an unusual shaking and trembling of the Earth and indeed of everything that before seemed to stand fixed upon it. Houses rocked like so many huge cradles, and in them tables, stools, trunks and chests rolled to and from with the violence of the shog; which put the people into so great a consternation, that they ran away and forsook their habitations … this Autumn Ague-fit of the Earth-Hag put us into a strange consternation all about Thame and Wallingford … a poor labouring man, a mean thrasher, being at his work in a small village two miles from Wallingford, felt this same shaking of the Earth, which he minded not at first; but when he heard the rafters of the barn begin to crack, away he ran, flung down his flail, and put the whole street in uproar.'[11]

I cannot prove that this earthquake shook all the trees in Lambridge Wood, but if the effects were so violent at nearby Wallingford it is almost certain that it did. Even the oldest woodland ticks off no more than a moment on the scale of geological epochs, and from time to time an ancient memory stored deep in the earth is rekindled. I expect that the fault whose revived movement caused the 'Ague-fit of the Earth-Hag' lay deep in the unconscious mind of southern England, in rocks contorted and shocked long before the age of dinosaurs and ammonites. The white chalk forming the backbone of the Chiltern Hills was laid down in the sea more than three hundred million years after the deeply buried rocks that hosted the fault. The 'Ague-fit' reminds me that each of the building blocks that make the landscape is ultimately shaped by globally shifting tectonic plates, and that these movements will continue their inconceivably slow and inexorable march regardless of the comings and goings of mere humanity with all his woods and all his works.

Nine years before the quake caused the thrasher to throw down his flail, the geological legacy of the Chiltern chalk had inspired a new industry. George Ravenscroft set up an experimental glass factory in Henley. He was attracted to the area because of the availability of particularly fine flint – that same flint that Neolithic hunters had appreciated. I added to the collection one flint nodule that I gathered from a chalk outcrop below the wood, a stone which bears a fortuitous and appropriate resemblance to the 'rother' of Rotherfield; inside, it is pure unspotted silica. Dr Robert Plot had come across the use of flint in glassmaking in his travels around Oxfordshire, so there was a tradition of flint glass in the area.[12] Henley would have been a comfortable place for George Ravenscroft to live because as a devout Catholic he could be assured of a welcome at Stonor House, where the Camoys had determinedly paid fines to maintain the practice of their faith.

The trouble with Ravenscroft's earlier efforts to make crystal glass was a tendency for it to break into a fine covering of cracks, known as 'crizzling'. The formula needed to be tweaked, but it was unclear how. Various compounds had been added to the melted flint to try to emulate the appearance of rock crystal. Eventually, replacement of lime compounds with just the right *soupçon* of lead oxide produced wonderful results. There is scholarly debate whether the mixture was discovered by Ravenscroft himself, or adopted from earlier experiments by Italian master craftsmen in Murano. However it originated, lead glass – 'crystal' – would go on to adorn all the best tables. The early pieces made by George Ravenscroft had the emblem of a raven's head impressed upon them,[13] so they are identifiable. I had thought a piece of this glass would make a fine addition to the wood collection. A lovely, squat, two-handled posset pot seemed a particularly desirable article. I soon discovered my mistake. If pieces ever come on to the market (which they very seldom do), they fetch a fortune. Most examples are in public collections like the Victoria and Albert Museum in London, where a visitor has the paradoxical experience of looking at glass – behind glass. Lonny van Ryswyck, who made a tile from the Grim's Dyke clay, was also able to melt our silica pebbles into a handsome green glass in one of her experiments, but it was not quite the same. I shall have to be content with my oddly shaped flint.

Winter is here at last. It is a clear day, and the wood somehow appears larger than usual. Television news has been full of stories of flooding. Pundits for and against man-made climate change have been exchanging words. One of the former implied that Chiltern beech trees might be threatened if the climate warms; nothing suggests warming today, but then this is weather, not climate. Icy wind whistles through the naked trees and I have to huddle inside my jacket. The frosty blue sky is riven with the trail of an aircraft – a white zip fastener across the heavens. It would not have been visible during the summer. A few branches are newly blown down, nearly every one scarred with the squirrel damage that has now become familiar. I see no sign of the culprits today – maybe they have withdrawn into the warmth of their drays. Maybe they are sharing a marmalade sandwich with Mr Mole. The damage could have been worse. A couple of tipped-up root plates remain in the wood where whole trees were blown over, though I cannot say whether it was by the hurricane of '87 or '90. Beech roots do not delve deep. One might imagine that this gale could bring another one down, but though there is squeaking and complaining from trees that are rubbing together, all the trunks are holding firm. The high branches toss chaotically. Something has been through the wood: deer have left abundant droppings like small black acorns, still glistening. I scoop up a few into a plastic pot (not for the collection). What can the roe deer have found to eat? I suddenly recall the phrase 'nipped in the bud' – and one of our planted hazels seems to have been topped, but nothing serious. Maybe a dog-walker sent the deer on their way, though only the hardiest, or foolhardiest, of that species would venture out in this weather. Grim's Dyke Wood has shut down until further notice.

What I take home with me today is the dung. Like a rotting log, dung provides a special habitat – in this case one supporting specially adapted, nitrogen-loving species, all of which are part of the biodiversity of the wood. Dung provides a demonstration in miniature of ecological succession, since one species follows upon another in a set order, like dignitaries in the Lord Mayor's Parade. It is much easier to see this at home than by repeated field visits. The droppings need to be prevented from drying out, but must not be wet. I find what works well is putting five or so fresh examples in a sealed transparent pot (the kind that olives are sold in) with some damped moss to keep the relative humidity high. Every few days the lid is removed for an inspection with a big magnifying glass.

Within a week, little white spindles drenched in water drops rise like rocket-launchers from each pellet. They look like germinating seeds, but they are the early stages of a remarkable mould called *Pilobolus*. A few days later tiny black spore capsules have developed, which look like minute 'hats' on top of the spindles. Many of these 'hats' have water droplets just beneath them. A special mechanism prompted by the droplets 'shoots' the capsules into the air – tiny black 'pepper grains' of spore packages decorate the lid of the container. In nature, shot spore capsules adhere to nearby vegetation and are nibbled by passing herbivores, eventually finding their way into their dung, where they germinate and perpetuate the species. The spore release of this 'hat thrower' fungus has acceleration speeds exceeding anything else in the natural world; but this is only the beginning of the parade. Small hairy white lumps now erupt from the sides of the pellet. Within a day or two these too have extended into stems, and transformed into conventional, if minute, mushrooms. Some are densely covered in white shining cells like snow (*Coprinopsis stercorea*); others look like miniature Japanese parasols (*Coprinellus* species). They are so delicate that a breath will destroy them, and their life is short, no more than a day. Their caps soon turn into a black mush – for they are tiny inkcaps, related to some species that occur commonly in the wood, like the magpie mushroom. Other droppings sprout little pink clubs of a weird mushroom relative called *Stilbum*. Now I wait for the cup fungi (Discomycetes) to appear: little yellow or orange discs, with a waxy look. *Lasiobolus macrotricha* is surrounded by long hairs, like eyelashes. There are finally some hard-to-see, black bomb-shaped fungi with hairy apertures that emerge from the droppings. Every time I incubate a sample I find something different. I intend to add 'incubating deer-droppings' to the list of pastimes in my biography.

Ice crystals grow from every twig.

9

December

Frost

It is completely still in the wood, sharply cold and invigorating, so that
every sense tingles with a special anticipation. All the puddles along
the tracks are topped with fresh ice, which has crystallised as blades.
When I tread on it, the ice breaks with a crunch like broken ginger
snaps, but there is still a little damp mud beneath it. A mist envelops
the trees: not a dense fog, but rather a pallid thickening of the atmosphere that increases with distance, turning the more distant tree trunks
into wan spectres. The bare winter branches above are a mesh of black
struts, but they too fade away in the deepening mist, and where the sky
should be there is only an airy nothingness. The whole wood is
cocooned in ethereal condensations that fence it off from the rest of the
world: there is the wood and only the wood. Nor is there any sound,
not even the hesitant cooing of a pigeon, or the distant scream of a red
kite. The absolute hush only increases my sensitivity to any chirrup or
the least rustle in the leaf litter, but apart from the crisp trudge of my
wellington boots there is nothing to disturb this enclosed world.

Every twig is decked with ice. From a distance it looks as if the
branches of each beech tree were spun sugar, making white arabesques
and swirls of impossible delicacy. Close examination reveals thousands of ice crystals sprouting from the thinner twigs in bunches,
making an attempt at icy replicas of the bundles of stamens that erupt
from them in springtime. Whole trees are clothed in white finery
more fragile than gossamer; the merest touch dislodges many little

bouquets of frozen rods that shatter into sparkling dust. Individual crystals are up to half an inch long, and every one is made from even smaller crystals stacked in line, so they have a fuzzy and complex profile. They have grown in the night, little by little, encouraged by the breathless stillness to become temporary stalactites built from dozens of minute motes of ice. Holly leaves have not escaped the attention of the invisible ice artist. Crystals fringe each leaf like a thin white ruff, and are spaced close together, but with surprising precision. The tips of the holly spines carry a single, tiny icy rod. Dry grasses dangle with diamonds that were made hours ago from thin air. Spider threads make necklaces for wood sprites. Even dry fallen beech leaves are dusted with tiny crystals like icing sugar.

I wonder if this morning is unique, or whether such brilliant decoration happens every few years if exactly the same atmospheric conditions are replicated. Maybe a woodward or a villein briefly paused here seven hundred years ago to admire similarly embellished trees, momentarily distracted from his hard life. Perhaps a Tudor politician was charmed for a moment or two from thoughts of intrigue at Court. Ice condenses from the mist, crystal by crystal, building on top of previous constructions. The miniature ice sculptures are breathed into existence from a haze that is as insubstantial as it is short-lived. A slight breeze would quickly undo all nature's delicate handiwork. If it breaks through, the sun will take an hour or two to destroy all the fine decorations. I will probably not enjoy such a privileged world again during my lifetime. *Sic transit gloria mundi.*

The holly and the ivy

'... When they are both full grown, of all the trees that are in the wood, the holly bears the crown'. As Christmas approaches, holly is the only tree in Grim's Dyke Wood that remains evergreen, other than the hardy yew. Whether ivy is counted as a tree depends on personal taste. It certainly makes wood, and can stand alone for a while if its support dies or is removed; the writer of the English Christmas carol evidently considered it tree material.

Holly (*Ilex aquifolia*) is the dominant understory tree in the wood. There might be too much of it. John Hill lives on the edge of Lambridge Wood, as he has for forty years. He cleared the holly in his part of the forest, and relates how it only came to be so common since management of the whole area was neglected in recent decades. Holly is a survivor. It flourishes even under beech canopy that captures almost all the summer sun. Its almost impenetrable interlocking growth provides us with privacy in the middle of the wood. The Dell around the King Tree is more or less encircled with hollies. Even in midwinter it is possible to sit there on makeshift stools of felled wood without being disturbed, to consume a sandwich or to make notes. On many overcast days old holly foliage looks almost black, and can appear forbidding. It is possible to push through it in spite of it bearing 'a prickle as sharp as any thorn', but going around is usually a more attractive option. Wood mice feel safe from owls under its close protection.

Holly does not grow as rapidly as beech, nor usually as high. One tree in the wood is comparatively huge, and has taken part in the race to the sky; it must be more than sixty feet tall. I assume that it germinated from seed immediately after the last big felling, and was able to compete with beech on equal terms within a generous clearing. Another substantial holly tree is 'twinned' with a big beech tree; that is to say, they are now growing side by side, and virtually married together, though they would have started as seedlings perhaps two feet apart. The beech has a girth of about eight feet, while the holly is a little over two feet around. I suppose that holly grows at very roughly one-quarter the rate of beech. Its evergreen habit means that a single holly leaf can photosynthesise for a long time, utilising low levels of light. The surfaces of the leaves are waxy and reflective, which is why they flash like silver jewels in bright sunlight.

Holly is a dogged kind of tree, not to be easily discouraged. Most of the trunks in the wood can be encircled by the fingertips of my two hands, and must be about twenty to thirty years old. They often form clusters of two or three trunks together, and shoot again from the base; and they sucker freely. I have pulled up suckers and found that they extend several paces from the parent tree – this is how holly extends its territory by stealth. I have been rooting out such rogue hollies where they might engulf my favourite areas, on the principle of never giving a sucker an even break. If a holly branch touches the ground it layers readily – it puts down roots and makes another plant. So this tough tree has a menu of strategies to outwit the tyro forester. The bark of holly differs from that of all the other trees in the wood: at least on younger individuals it is irregularly striped green along its length. Maybe it can even squeeze in a little extra photosynthesis from this unexpected quarter.

As might be expected from such a slow-growing tree, holly timber is unusually hard, which accounts for its use in turnery, marquetry and printing blocks – 'sturdy uses', as John Evelyn put it. Traditional craftsmen still appreciate its virtues. Once the bark has been stripped the wood beneath is remarkably pale – almost white. Clipped holly has long had employment as a hedge to deter pilferers and house-breakers. It does not frighten birds, which love to hop in and out of its generous cover. Holly suffers from few diseases, although I do not relish the thought of its gloomy forests taking over if other, more vulnerable trees die out. It only hosts one insect, whose larva makes a meal of a leaf by getting under its tough cuticle – the holly leaf miner fly (*Phytomyza ilicis*). Its presence in the wood is betrayed by vivid yellow and red patches that stain some holly leaves while they still hang on the tree.

Although its spikes render adult holly foliage intimidating, freshly emerged leaves are soft and afford nutritious fodder; before their protection hardens up I have crumpled them in my hand and felt no pain. Young branches were once cut as feed for stock by cowherds; if a deer discovers an emerging shoot it is nibbled at once. Erasmus Darwin noticed in *Phytologia* (1800) that prickles only develop on the lower leaves where they will afford the tree some protection: holly foliage on the tops of trees is completely without spines – it could be mistaken as belonging to a different species. *Ilex* provides a rather blatant example of adaptation, and who can say whether it was one of many ingredients in the intellectual *potpourri* that encouraged

Erasmus's grandson, the young Charles Darwin, towards evolutionary theory. Holly flowers are small and white with four petals, discreet but pretty enough, tucked among the leaves at the end of the shoots in May, with separate sexes on different trees. Although both flowers look superficially similar, males are readily recognisable when they shed masses of yellow pollen on to the leaves below them. The glory of holly, and the reason for its incorporation in the Christian tradition, is the berries that develop on the mother trees, 'as red as any blood' when they colour up late in the year. I had kept an eye on the green, unripe berries, waiting eagerly for this transformation. At this point I must confess to one of my failures in the Grim's Dyke Wood project.

The Millennium Seed Bank is a bright, modern construction in the grounds of Wakehurst Place, Sussex, which is the 'country seat' of the Royal Botanical Gardens, Kew. Walking around the perimeter of the new building, I glance through glass panels at white-coated scientists in spotless laboratories. Like any other bank, this one protects capital, but the wealth here is the currency of the future of the home planet: a huge collection of plant seeds, carefully dried and then stored at minus 22 degrees Celsius in rank after rank of specially designed 'safes'. This capital will last for years: it is a Fort Knox of biodiversity. Many of the plants whose genetic blueprints are safeguarded in these frigid cabinets are rare in the wild, so the Wakehurst building is also a kind of Ark.

I was there to be instructed in how to contribute to a different project. In order to assess the genetic diversity of our woodlands, collections of tree seeds are being made by naturalists throughout Britain, then processed by the staff at the Bank. Quite apart from long-term conservation, the ability of trees to defend themselves against new diseases relies on having a pool of genetic variation to draw upon, and the Seed Bank archive might prove crucial in decades to come. Most of the trees described in this book are wind-pollinated, and that means their genomes can readily spread: these collections will reveal patterns in genetic distribution that will help predict what will happen as climate changes.

The protocols for collecting seeds are quite strict: they have to be fully mature to freeze successfully. For holly, the ripe berry has to develop what is called an 'abscission layer' from the parent plant – a natural seal – so that it breaks off easily. Even though each berry yields four seeds, I am supposed to gather thousands of seeds per sample, so I know it will be hard work. The birds in the wood, however, had other ideas. When holly berries are ready for picking for science they are also on the menu in the wild. During a cold winter like this one they will be sought out enthusiastically. Our own blackbirds are greedy enough, but when redwings arrive on migration from Scandinavia the berry crop is doomed.* By the time I realise what is happening it is already too late. I begin to appreciate how so many holly seedlings come to be dotted around the wood. They were the end products of excretion from a fully sated member of the appropriately named Family Turdidae (blackbirds, thrushes, redwings, fieldfares, etc.). Next year, beechnuts will make a simpler target.

As for ivy (*Hedera helix*), its purple-black berries are not quite ripe for Christmas. They are held in bunches away from the foliage, like small bouquets of plump black peppercorns. They are more conspicuous than the clusters of green flowers in October that preceded them, although their modest blooms attracted many wild bees. In the wood ivy flowers and berries are only produced high off the ground, but in a nearby hedgerow I have heard an ivy bush humming from the attention of its enthusiastic pollinators. Ivy grows to the height of the tree up which it climbs, sticking to its host by means of adhesive brownish 'rootlets' that emerge from the underside of the fast-growing shoots. I find the young, arrow-shaped leaves rather beautiful as they alternate left and right up the plant, each leaf decorated with a scaffold of a few pale veins. Mature, five-pointed, dark-green leaves borne on slender stalks are efficient photosynthesisers like those of holly, which is why ivy can spread over the darkest ground seeking out new hosts to climb. Leaves carried high on a growing vine also change their

* We finally succeeded in beating them to it in 2015.

shape in a similar fashion to holly: those on the flowering shoots are often no more than simple ovals with a terminal point.

In woods where every tree is choked with ivy the effect can be oppressive, as the poor trees struggle to cope with their evergreen burdens, but in Grim's Dyke Wood only two beeches carry enough ivy to obscure their trunks, and I welcome them as the only wild vine I have. Jenny wrens like to build their nests in the security of the thick cover. Wood pigeons will eat the berries when winter really begins to bite. In the spirit of experiment I nibble a berry, but do not swallow it. It is extraordinarily bitter. I suspect that avian tastebuds work differently from mine.

Holly and hazel make the best walking sticks. I was given a book all about the craft of making sticks, full of magnificent examples way above my whittling skills, featuring handles carved into foxes' heads and silver trimmings. However, it is perfectly possible to make a serviceable walking stick even with my impoverished DIY talent. It is a question of choosing the right standing stick in the wood. Women's sticks can be more delicate than ones intended for men, but both should be as straight as possible. This is easy with hazel poles. Very straight holly sticks are harder to find, and usually require several little branches to be trimmed off close. I detected the work of a 'phantom stick-cutter' taking a few of my holly sticks in the wood, betrayed by the neat cut made by a collapsible pruning saw. Good luck to him. The cunning plan for the easiest product is to cut a stick with a handle made by nature. Hollies often have a crook in them, which is a handle waiting to happen. Hazels cut near the base can thicken into a usable club-handle.

I dried a few holly sticks for a year before using them. My first attempt was rather thickset. Jackie described it as a 'cudgel', which I thought was cruel. A second attempt removed the green bark with a knife except around the natural handle. Much polishing with sandpaper was probably an inefficient way to produce a smooth white stick – a coarse rasp had to help eliminate the 'knots' where branches attached – but the end product was quite handsome. The same rasp and sandpaper finished off any sticky-out bits around the handle.

When polished up, a straight holly stick looks very much as if it were intended for the use of a blind person. For those who are not blind, it is probably wise to stain the stick using a strong solution of instant coffee, which turns the wood a lovely golden colour and highlights the grain. The addition of a rubber ferrule bought from one of those hardware shops that are *not* DIY superstores completes this rustic, though efficient walking stick. The stick has to squeeze really tightly into the ferrule, otherwise you will lose it on your first strenuous walk.

Slaves and highwaymen

Another, rather large portrait hangs in the River and Rowing Museum, painted by the Dutch artist Jan Siberechts as the seventeenth century drew to a close. This charming portrayal of Henley-on-Thames shows the town nestling (as the brochures always describe it) in the Chiltern Hills by a working River Thames, the old flashlock conspicuously in the foreground. The tower of St Mary's church dominates the town, as it still does today. Cleared fields occupy the lower ground; abundant woods – including I dare say Lambridge Wood – clothe the more distant hills. Piled adjacent to the wharves are stacks of timber products waiting for export to London. Other paintings in the same series show covered goods stacked on square-ended barges ready to leave town, and a range of commercial properties close to the river, which would prosper still further in the century to come.

Siberechts's competent panoramas are reminiscent of contemporary paintings of Netherlandish towns exuding an air of understated prosperity. Simon Townley has drawn attention to the number of smoking chimneys in one of the townscapes, proving that small interior fireplaces had increased in general use, for bourgeois comfort rather than for show or for cooking. Window glass was no longer a luxury. Local wills and probate inventories testify to increasing affluence during the next decades. Henley's importance as a trading centre was undiminished, as was the part that woodland played in the business of business. In 1726 Daniel Defoe wrote of the woods just downriver in adjacent Buckinghamshire: 'Here is also brought down a vast quantity of beech wood, which grows in the woods ... more plentifully than in any other part of England. This is the most useful wood, for some uses, that grows, and without which, the City of London would be put to more difficulty, than from any thing of its kind in the nation.' Defoe tells us that its principal uses were for 'fellies [felloes, p.136[1]] for the great carrs ... which ply in London streets for carrying of merchandizes', 'for billet wood for the king's palaces, and for the plate and flint glass houses', 'for divers uses, particularly chairmakers and turnery wares'. 'The quantity of this brought from hence is almost incredible, yet so is the country overgrown with beech in those parts, that it is bought very reasonable, nor is there like to be any scarcity of it for time to come.'[2] Woodland continued to be an investment that constantly renewed itself: for now, Grim's Dyke Wood was safe from grubbing. Of Henley itself Defoe noted particularly 'the trade from malt and meal and timber for London, which they ship, or load, on their great barges'.

The town was due for a facelift, a description that might be exactly right. Fresh faces were added to the fronts of older buildings, their medieval bones dressed in new finery, carrying further a process that had already begun in the seventeenth century. Several eighteenth-century houses now boasted double-fronted classical proportions, larger windows with more generous panes as glass manufacture improved,[3] maybe a fanlight above the central doorway, which was flanked by pillars. Nettlebed bricks came into their own to make town houses with chequered fronts. The tanner vied with the brewer to dress up his medieval mutton as Enlightenment lamb. What Italians refer to as *la bella figura* may have set the 'look' that Hart Street and Duke Street in Henley still carry today. The ancient realities of burgage plots and oak frames were still lurking behind the newer façades. I trust they will still be there when the Starbucks signs come down and when the chain stores go bust: deep history has a propensity for survival.

The owners of Greys Court – and of the wood – were changing again. In 1724 the title to the manor had passed from the last of the Knollys line, by virtue of marriage, to Sir William Stapleton, Fourth

Baronet. The new owner was descended from a William Stapleton who was largely a self-made man, and rose to become Governor General of the Leeward Islands in 1671. The first William grew wealthy through owning sugar plantations on four of these West Indian islands (St Kitts, Nevis, Antigua, Montserrat). His money and status as First Baronet was achieved on the back of slavery, which was usual when British territories in the Caribbean were colonised and 'developed'. He appears to have been a hardworking and trustworthy administrator by the standards of the time. It was not uncommon for descendants of moneymakers to remove themselves from the relentless toil and unhealthy life in the colonies, while still enjoying the fruits of the labours of people they did not care about. The Fourth Baronet was an absentee slave-owner who aspired to join the gentry of Oxfordshire. He tried, nonetheless, to extract as much money as he could from the portion of the family estates he still owned on Nevis. His manuscript letter-book is preserved in the Harvard Library,[4] and shows William badgering his agent Tim Tyrell to get what he regarded as his due from the plantations. He was convinced that he was being cheated, that the expenses of running the plantation were too high, that the estate managers could not be trusted; but the daunting voyage across the Atlantic Ocean to check up on matters in person was out of the question. If anything, he was 'running' the plantation on too few slaves, of whose conditions he knew little.[5] In 1725 Tyrell had reported to William's mother 'a great many negroes that died from want of provisions as well as by sickness ... ye caterpillars and worms have swept away a great many young canes for ye next year's crop'.

The unappealing picture of Sir William when acting by proxy in Nevis is reinforced by his reputation at home. He was elected to Parliament for Oxford unopposed in the general election of 1727, but as Dr Stratford of Christ Church College remarked: 'Our knighthood of the shire went a-begging ... One Sir William Stapleton, a West Indian, formerly of Christ Church, a rake then and I hear he is still, is to be [the] man.' Once elected, he voted against the administration in every division on record, and made but one speech, in 1733, when he successfully opposed rum being imported from the North American

colonies into Ireland because of the damage it might do to his plantation interests in the West Indies.

William's son Thomas was born at Greys Court in 1727, but was
not resident there for some time after his father's death in 1739 (it was
let to the local parson). By the time he took up residence again at the
age of twenty-four he was well on his way to becoming something of
a rake in his father's mould. His cousin Francis Dashwood lived not
many miles to the north across the hills at West Wycombe. Young
Thomas enthusiastically joined the Medmenham Monks, aka the
Order of the Friars of St Francis of Wycombe, today usually recognised as the notorious Hellfire Club. The caves where this group of
free-thinking libertines allegedly got up to all sorts of tricks are now
a tourist attraction opposite Dashwood's house next to the River
Wye.[6] Meetings of the 'Friars' were held at Medmenham Abbey on the
Thames between Henley and Marlow. Since no detailed minutes were
kept it is not clear whether the posthumous reputation of the club has
been embellished, and it has even been suggested that it had political
intentions, but it is certain that its members were mightily against
religion, and mightily for individual freedom. A Hogarth portrait of
Dashwood as St Francis portrays him reading an erotic novel where
the Bible should be. Alcohol was liberally involved. The National
Trust guide to Greys Court rather proudly tells readers that in 1762
'Thomas de Grey [a nickname used by Stapleton] and John of Henley
consumed four bottles of port, two of claret and one of Lisbon at one
sitting.' Claims that the 'Friars' met in the dower house at Greys are
consistent with a Latin inscription there that insists 'Nothing is better
than the bachelor life.' The Hellfire Club dotted its other venues with
Latin tags.

The irony of the Fifth Baronet as champion of individual liberty
deriving some of his wealth from slavery does not have to be dwelt
upon. A further irony is the use of de Grey as a nickname, since that
too returns us to a previous age when near-slavery was part of the
feudal system. Thomas Stapleton did promise to reform on his
marriage in 1765 to Mary Fane, who hailed from yet another nearby
Chiltern estate at Wormsley. By now, connections between the

grander families were established all over the southern lobe of Oxfordshire and the adjacent counties. Thomas added a delightful floral plasterwork ceiling to the drawing room of the big house to welcome his new wife, and extended the east end of the building with a typical Georgian bay that would not look out of place in Mayfair. He also built the ha-ha that continues to enhance the view from the lawn. Quite soon, he ran short of money. He could not afford wholesale rebuilding, nor to construct grand parks like those that would be planned nearby. For our wood, it meant continuation of its old ways, and another escape from obliteration. But now it was connected to a world so much wider than in the time of the de Greys, when the estate was nearly self-sufficient; for the economy of which Grim's Dyke Wood was a tiny part stretched halfway across the world to the Leeward Isles.

Fawley Court: engraving after a drawing by J.P. Neale, c.1826.

On the other side of the Fair Mile, the Fawley Court estate was bought by William Freeman, who completely rebuilt the house in 1684, after the ruination it had suffered in the time of Sir Bulstrode Whitelocke. Freeman was another 'West Indian', born on St Kitts in

1645, who made his fortune as a sugar merchant in London, from the proceeds of which he purchased his Henley estate. He also retained estates in Nevis, which he continued to manage as an absentee planter. His copybook shows him taking some interest in the welfare and nutrition of his slaves, even training them up for 'trades' like cooperage. It is extraordinary that Stapletons and Freemans were neighbours both in England and on the other side of the Atlantic Ocean. Every quarter that could be seen from our wood at the end of the seventeenth century was partly underwritten by the profits from slavery.

The Freemans who followed in the eighteenth century prospered greatly in business and politics. The grounds of Fawley Court and Henley Park came to be regarded by their wealthy owners as so much landscape that could be moulded to aesthetic ends. From their side of the Fair Mile, Lambridge Wood played a new role – it became a distant prospect, a component element of the view. John Freeman built a 'tumulus' in Henley Park, under which he buried various bits and bobs of broken crockery and household items as a 'time capsule' that might be discovered by future archaeologists.[7] He was an early member of the Society of Antiquaries. His son Sambrooke Freeman was more ambitious, remodelling Fawley Court in the latest neoclassical style, and between 1764 and 1766 employing the most prestigious of landscape architects, Lancelot 'Capability' Brown, to design the grounds. Sambrooke commissioned one of Henley's famous landmarks, the classical 'temple' on Temple Island in the middle of the river at what is now the start of the Regatta course. The folly was used for fishing and picnics; the sophisticated gentry who dined there admired the scenic qualities of river, hills and woods all around. Sambrooke Freeman appears to have considered much of the Henley area as an extension of his park. He purchased more land and manors as part of the project, including Phyllis Court on the north side of town.

When at last in 1781 the ramshackle wooden bridge over the Thames was replaced, the aesthetic qualities of its stone successor were much discussed by the local grandees; after all, it was going to be an important part of their prospect. General Henry Seymour

Early-nineteenth-century view of Henley Bridge from the bottom
of New Street, with a barge tied up at the wharf.

Conway (Commander in Chief of the Forces) himself attended the
planning meetings, crossing over from the Berkshire side of the river
where he owned Park Place, yet another grand house set in its own
extensive landscaped grounds. It must be admitted that the 'gentlemen' had excellent taste: the present bridge is a pleasing and elegant
construction that sits wonderfully well just where it is. General
Conway's daughter Anne Damer designed the stone faces portraying
Thamesis and Isis that still overlook the flowing waters. This Georgian
elite could be enlightened enough to conceive of women as artists.
The social structure of the Henley countryside was by now established: it was 'posh'. The landscape was for the enjoyment of the
wealthy. That exclusive image has stuck.

The exploitation of woodland continued at Stonor, where income
from the estate still guaranteed the future of a family that had owned
the same land in the Assendon Valley for centuries. The papers and
records of the estate provide an unmatched picture of local silvi-

culture. In the middle of the eighteenth century Thomas Stonor advised his uncle Talbot on how to manage the woods while he was absent. 'The woodmen order the workmen what trees are fite to be felled and when a wood is felled sufficiently. They are to give notice to the steward that he or his Deputy take the tale [tally] thereof and immediately pay the workmen for the same.' Other obligations were to be discharged: 'a woodfeast at my expense at Henley to encourage the chapmen [middlemen] to bring part of their money ... within a month after midsummer'. There were instructions on how to prevent workmen stealing 'shouldersticks', but allowing them the odd 'brush fagot'. Hedging and ditching were necessary to define areas of coppice, and to keep deer from the regenerating growth. William Strongharm's disbursements in 1749 include £10.3*s* 'for heging and diching' – could a better name be imagined for an estate employee than 'Strongharm'?

The account books prove that oak timber was more valuable than beech. Hundreds of loads of firewood embarked at Henley, where wharfage was paid. Bavins for bakers, town billets for hearths, long faggots for kilns; nothing was wasted. Tenants were allowed 'wood-boote', 'cart boote' and 'hurdle boote'; the rights of 'botes' dating from the Middle Ages still continued. There is a timeless quality about these account books, a sense of centuries recycled, and yet new markets were opening up. Wood for furniture is listed, as Daniel Defoe had noticed in Buckinghamshire. A new use for beech was in lining the canals that were feeding the nascent Industrial Revolution. There is no way of knowing how many of the practices up at Stonor were applied in Lambridge Wood, but it is certain that our woods too were managed for income for the Stapletons. In due season, men with names like Strongharm and skills to match tramped through the beech trees about their business.

Stonor and the other Chiltern Hills estates could prosper sufficiently on a mixture of arable and, to a lesser extent, pastoral farming, combined with exploitation of their 'semi-natural' ancient woods. A rather wonderful continuity was maintained. During the eighteenth century the relative proportion of wheat increased at the expense of other grains in response to demand from London.[8] Methods of soil 'improvement' and crop rotation, coupled with mechanical sowing and harvesting aids like those championed by Jethro Tull, led to more efficient farming and higher yields. Despite these changes, ancient patterns of land use continued in South Oxfordshire, while they progressively disappeared elsewhere. In the hills, such enclosure of fields as was necessary had already happened long ago.

Meanwhile, on the flatter territory beyond the Chiltern scarp the countryside was undergoing a profound reorganisation. The Vale of Aylesbury and all the low ground towards Oxford was starting to resemble the pattern of generous fields, so often surrounded by thorn hedges, that we now think of as 'typical' lowland English countryside. The medieval arrangement of open fields, strips and common land was inexorably transformed into the planned patchwork of fields that look so appealing when viewed from Watlington Hill, or from an aircraft taking a wide loop around London.⁹ 'Inclosure Acts' were passed in Parliament that legitimised a much more efficient use of fertile land. The same Acts caused considerable grief to the ordinary folk in rural communities, as they were poorly compensated for the loss of their traditional rights. However, Chiltern Hills life continued in many ways much as it always had done. The mix of woods and fields around gentle dry valleys that so delighted Sir Sambrooke Freeman and his friends was an ancient landscape that persisted despite advances in agricultural techniques. Chiltern woods are survivors, including our own. The whole area is a 'time capsule' more authentic than that devised by the antiquarian John Freeman at Fawley Court.

The road running past our wood over the Chiltern Hills to Nettlebed and Oxford has already figured several times in this history. If eighteenth-century country estates were affluent and ordered, the roads connecting them were appalling. Here is Robert Phillips's description in 1736: 'In the Summer the Roads are suffocated and smothered with Dust; and towards the Winter between wet and dry, there are deep Ruts full of Water with hard dry Ridges, which make it difficult for Passengers to cross by one another without overturning; and in the

Winter they are all Mud, which rises, spues, and squeezes in the Ditches; so that the Ditches and Roads are full of Mud and Dirt all alike.'[10] Negotiating hills was even worse. The old road up the hill beyond the Fair Mile to Bix at the top is still open, running past Cecil Roberts's Pilgrim Cottage into a deep and dark holloway that was cut by countless carts as they laboured up the steep chalk slope. Drovers had probably followed the same route centuries earlier. Now the old track's brooding presence is enhanced by neglect: tree roots twist out of the high banks, and hazels that should have been coppiced years ago lean out from both sides to almost meet across the middle, turning the road into a tunnel. Ivy runs amok, and dangles in festoons from overhanging branches. It is a place of deep memories, where the past might stage an ambush at any moment.

Three hundred years ago the chance of ambush was real enough; travellers would have more to worry about than mud or a broken axle. As the road from Henley passed up into the woods there was cover and opportunity for robbers of all kinds: 'If you beat a bush, 'tis odds you'll start a thief,' as a local saying had it. Lybbe Powys of Hardwick House[11] kept a diary detailing the almost relentless social whirl among the affluent families around eighteenth-century Henley – one long round of parties, plays and lavish suppers. In 1777 she had a narrow escape. 'Miss Pratt and I thought ourselves amazingly lucky' to get away unharmed when the chaise behind theirs was robbed. 'It would be silly to have lost one's diamonds so totally unexpected; and diamonds it seems they came after. More in number than mine indeed.' On 19 December 1779, 'Mr. Powys and Tom went to Bletchingdon Park to shoot, and were robbed by a highwayman only four miles from Henley on the Oxford road, just at three o'clock. We hear the poor man was drowned the week after, by trying to escape (after having robbed a carriage) through some water which was very deep. He behaved civilly, and seemed, as he said, greatly distress'd.' The robbery must have happened just beyond our wood, towards Nettlebed.

The sad story of this particular crime was not unusual; some gentlemanly highwaymen were driven to desperate measures by desperate circumstances. Isaac Darkin, however, was of the swash-buckling kind that would have been played by Errol Flynn in an old-style Hollywood movie. His weakness was the fair sex, and he was an attractive rogue. As the Newgate Calendar said: 'Darkin was so distinguished by the gracefulness of his person that he was the favour-ite of unthinking women wherever he came.' His adventures as a gentleman of the road led to an early deportation to Antigua, where he found army service not to his taste. By devious means he escaped back to England, where he took up his old ways with gusto. His undo-ing came about when 'he stopped a gentleman named Gammon, near Nettlebed, and robbed him of his watch and money'. Darkin could very well have passed through our wood on the way to his last robbery, but on this occasion he could not wriggle out of responsibility for the crime. He was arrested while in bed with a 'woman of the town'. He treated his death sentence insouciantly, and the Calendar tells us: 'On the day of his execution his behaviour was remarkably intrepid; and at the place of his death he fitted the halter to his own neck.' He was twenty years old.

Something had to be done, both about the roads and about the foot-pads and highwaymen. Less than half a mile from Lambridge Wood at the top of the hill at Bix, an oddly angled single-storeyed white house stands adjacent to the highway. The windows at the front of the little building face both ways along the road, so the occupier could spy on who was coming and who was going. This tollhouse for the turnpike is a legacy of the road improvements that slowly transformed transport around Britain. Turnpike Trusts were established to improve sections of road – at a profit for the trustees. Tollhouses were quickly thrown up along many of the main routes; at least three survive around Henley. The practice of preventing passage of vehicles that had not paid their fees by putting a pikestaff across the road did not last long, but it did give us the word turnpike, which now applies to highways in the USA, which are about as different from the road from Henley to Wallingford by way of Nettlebed as could be imagined. I count driving down (or was it up?) the multifarious New Jersey Turnpike in rush hour as one of the more terrifying experiences in my life.

Late-eighteenth-century tollhouse at Bix.

The old road below our wood was turnpiked in 1736, but followed the traditional route, which can still be walked today; beyond Bix it traces the valley bottom towards Nettlebed, and becomes very boggy very quickly in wet weather – four-wheel-drive vehicles from local farms struggle along it. I cannot imagine that the route was greatly improved when the tollhouse was set up in 1772. The new roads were far from popular. As Cecil Roberts describes it: 'Disturbances and protest meetings broke out all over the country against the extortions of the Turnpike Trusts. Soldiers had to be called out. Two men were hanged at Worcester, and one, after being hanged at Tyburn, threw back the coffin lid as it was being screwed down.'[12]

Only the wealthy could afford to travel long distances. Towards the end of the eighteenth century a brand-new, straight road from the Fair Mile to Bix was cut along the same line the main road follows today, leaving the old road up the hill to dream in its ancient holloway. Later, another new, direct road was cleared through the woods to Nettlebed. Road surfaces of crushed chalk were hardly adequate, and were replaced by firmer gravels; drainage was improved, so that wash-outs became less frequent. Secure roads with better oversight meant that highwaymen could be pursued more effectively and brought before the justices of the peace. This has been attributed to the pursuit of offenders by 'the Steward of the Chiltern Hundreds [which] was not the sinecure that it is today'.[13] This particular Crown office dates from medieval times, but is now awarded to Members of Parliament who wish to resign, as a ruse to allow them to leave public service – the law makes it impossible to hold two Crown appointments at the same time. The House of Commons Information Office tells us that the seventeenth century was already 'a hundred years after any records of [the Stewards'] administration cease'. So it is likely that the demise of the highwayman was more a matter of local vigilance and better roads. By the nineteenth century there were still poachers in our woods, but the pistol-toting rogues who had haunted the route across the Chiltern Hills had vanished forever. The regular clatter of the hooves of horses pulling post chaises would now carry up the valley sides to mix with the soughing of the branches in the wind and the cawing of surprised crows up among the lichens.

Lichens

No space available for life is wasted. A high wind has been through the wood overnight, and has brought down twigs from the highest boughs of the trees. All the twigs are decorated. I have only to pick them up from the ground, for the wind has done my sampling for me. Some carry bright-green patches, and others greyish, crimped crusts resembling frayed paper. One piece of ash is covered with golden-yellow wrinkled decoration: a series of discs fixed like imperfectly finished appliqué, rougher and more orange-hued in the centre of the highly irregular patches. A cherry twig carries what looks much like bunches of seaweed that have dried a little after the tide has gone out.

All this natural bunting is lichen, which likes to grow in the wetter, winter half of the year. More than a dozen kinds grace the wood, and when they are dry they are readily added to the collection. Leafy-looking lichens live on the higher branches near Grim's Dyke, although I have found them on low shrubs along the track into the wood where there is more light, and particularly on elderberry, which they seem to love. They apparently feed on thin air, because there is no soil for them on twigs, nor do they steal anything from their hosts, other than a place to sit. They dine on dust, and drink rainwater. Lichens are a close collaboration between a fungus and a photosynthesising partner – usually an alga or blue-green bacterium (sometimes both) – so they provide another example of symbiosis. It is a strange thought that such cooperative behaviour is so important at both ends of the tree: in the roots as mycorrhiza, and up aloft in lichen-decked branches. Lichens dress the bare twigs, dappling here, frilling there. The algal partner steals its nourishment from the sun, and the fungal threads that wind between and around the plant cells provide other nutrients. It is hardly surprising that lichens grow so slowly, but they can afford to take their time. They endure drought more effectively than almost any other living organism. Water brings them back to life, and when there are no leaves on the trees, there is additional light, so these inconspicuous piggybackers prosper while other species go into hibernation.

I examine the bright-yellow lichen with my hand lens to discover groups of tiny, intensely orange cup-like structures towards the middle of the patch. On the gold shield lichen (*Xanthoria parietina*) the minute spores of the fungal partner are borne in these cups.[14] The yellow pigment provided by the fungus protects the algal partner from ultraviolet radiation. It's a real *quid pro quo*. A smaller grey, finely branched and crimped lichen makes less defined masses further along the same twig. It is not surprising to find *Physcia tenella* near the gold shield lichen, because both can tolerate high nitrogen levels. A grey, leafy shield lichen (*Parmelia sulcata*) also lives among them. These species can regularly be found together in urban environments.

Lichens are nature's chemical bellwethers. Their unrelentingly exposed lives means that they incorporate whatever the atmosphere can throw at them. Nitrogen is concentrated by excessive use of fertilisers in modern farming and by pollution in cities. Many lichens cannot cope with this element at all – now, such choosy species tend to live in the west of Britain, where the pure sea air keeps them happy. When highwaymen lurked in the woods the same species would have been comfortable growing in the Chiltern Hills. Subtle changes in the atmosphere do not register with human nostrils, but lichens cannot be fooled; they can taste micrograms, and they can smell molecules. Mankind's influence infuses the air itself.

Other lichen species in the wood point to recent improvements in the environment, so it is not all bad news. Glossy grey-green leafy lichens called *Punctelia subreducta* may have reappeared because of reductions in levels of sulphur dioxide. *Usnea subfloridana* produces dangling bundles of fine, grey-green whiskers very sensitive to the same industrial pollutant; this lichen is reminiscent of 'Spanish moss' that hangs from almost every tree in the Florida Everglades (though it is botanically unrelated). The appearance of both of these lichens may well reflect the great reduction in coal-burning over the last fifty years – in other words, 'cleaner' air.

Only a few lichens, like the fresh-tinted greenshield (*Flavoparmelia caperata*), are worldwide in distribution, streetwise survivors in any situation. On our wild cherry twigs oakmoss (*Evernia prunastri*) sprouts miniature, grey-greenish 'shrubs' made of dense clusters of forking shoots. This species has been harvested as a perfume base, and used as an 'antiseptic, demulcent, expectorant, and restorative' (i.e. cure-all) by herbalists. Even apparently smooth beech branches carry inconspicuous greyish crusts that betray their lichen affinities when their little reproductive cups – blackish rather than orange – are identified under a lens (*Amandinea, Arthonia* and *Lecania* species). Lichens, in short, thrive just about everywhere in places where other life is daunted. They also make a mini-ecosystem of their own: three species of footman moths trapped by Andrew Padmore use lichens as larval foodstuff. And they are inextricably linked into the wider world: into subtle changes in the atmosphere, into agricultural practices, or even political decisions about energy supply made by people in offices who rarely venture into woods. By Georgian times, trade routes connected the owners of our wood to the New World, and new turnpikes linked it more closely to London and Oxford, while high in the beech and cherry trees lichens were already there, as perpetual sentinels, alert to invisible changes still to come.

10

January

Second felling

It is eight o'clock on an unseasonably mild morning, and dense mist pervades the trees; the open fields beyond the wood have apparently vanished. A discomfited buzzard flaps away into invisibility, with a scream. Martin Drew arrives in his Range Rover, ready to take down two wild cherry trees around the big clearing. One is at its peak for harvesting: a fine tree vying for space with the beeches, and as tall as any of its rivals. We had to get permission for felling from Natural England. The other is the sickly tree I had noticed earlier in the year: its sparse leaves had fallen early, and there are signs of fungal infection under the bark. Martin tells me that its heartwood will still be fine.

The best trajectory for both trees when they fall is a matter for an experienced woodsman. Martin is a man of few words but careful contemplation. He stares intently at the doomed cherries, making calculations. He decides to fell the sickly tree first. The back of his vehicle hides several serious chainsaws, and exactly how and where these cut into the base of the tree determines where it comes down. The buzzing and grating saw moans and complains as it gets to work, silencing or outcompeting any songbirds that might have been fooled into thinking that spring had come early on this oddly warm day. A mighty crack and then the cherry tree falls down in a few seconds, collapsing across the public footpath. It is as well Jackie and I had acted as guards to keep walkers clear. As it falls, the trunk effortlessly breaks off a substantial beech branch in its way – quite cleanly, like an amputation. Now it is the turn of the big tree, all eighty feet of it, and most of it polished trunk, straight and noble. It must contain a good quantity of fine timber. Martin slices obliquely at its base to try to direct its fall into the open part of the clearing. It would be difficult to extract if it got snagged with any of its neighbours. The grinding noise of the chainsaw goes on and on, like those dentists' drills you fervently wish would stop. Then there is a sudden wrench and a crack, and the felled tree is down perfectly and in place: there is hardly time to photograph the event. The leafless high branches derived from its crown break and settle down with the briefest of sighs. The end of decades of photosynthesis is recorded with a bang and something of a whimper.

Martin's silent colleague has arrived, driving a huge tractor that pulls a cradle. The machine is equipped with a powerful mechanical grabbing arm. The smaller branches – many of them as thick as my leg – are cut off from the main trunk with the chainsaw; it is a matter of a few seconds per slice. It is hard to dislodge an image of the deliberate dismemberment of an animal. The twigs are richly covered in bright-green and greyish lichens that only a few minutes earlier were open to the sky; it was a garden up there. Next, the trunks themselves are sliced into three massive pieces. The grabber lifts each length with ease, and swings them up into the cradle. I am reminded of a fairground game that we used to play for a few coins; my children always wanted to win the big prize when a miniature grabbing crane swung into action, dipping and clutching over a pile of toys and gaudy watches. (Nobody ever won the big prize.) The process of gathering our timber seems as effortless. The two trees don't even fill the cradle.

The stumps remain behind. With their buttresses sliced through, the bases resemble huge starfish stranded on the floor of Grim's Dyke Wood. The heartwood is a wonderfully warm pink colour, almost like boiled ham, while the sapwood tends towards orange; the fleshly comparisons remain ineradicable. Some of the offcuts will provide us with firewood for a year or more. After they have been cut into rounds I will split them at home with my beetle and wedge into usable logs. The thinner twigs will decay back into the soil in the wood.

Jackie and I follow the trailer at a dignified pace to Martin's sawmill at Culham, on the plains beyond the Chiltern Hills. This time our felled trees are going to be used for something other than firewood. The sawmill is tucked away behind the European School, and in sight of the remaining cooling towers of the Didcot Power Station, with large fields beyond; such a different landscape from the wooded flanks of the hills where the cherries grew to their lofty maturity. The yard, however, is just as it should be. It is piled high with tree trunks in every state: gnarled old oaks stacked higgledy-piggledy alongside beech destined to become firewood; hoppers spilling over with logs for burning. Rank upon rank of previously planked trunks are arranged around the yard in batches for future clients to convert into floors or furniture. The best length of our finest cherry is sliced by a travelling bandsaw with the ease of a cheese wire cutting through a round of Cheddar (I notice the equipment is labelled 'Tom Sawyer'). The cutting grind of the saw is certainly deafening, but lacks the complaining tone of the chainsaw. Martin's earmuffs protect him from the worst of it. Sawdust has blown into a dune downwind of the contraption. Martin rides his machine as would a bus conductor the platform of an old London double-decker bus. The finest log can be cut into the thinnest planks – a mere 30–40 millimetres at best – but several thicknesses are extracted from the same length of timber. Martin controls the thickness by twiddling on a gauge at each cut. After they are sawn, the planks finish up lying on top of one another like a stack of playing cards. The sick cherry does indeed have intact heartwood, but it was pocked and slimy under the bark; it would not have survived another season. Fine sawdust blows into my eyes, and it is time to retreat.

After planking comes seasoning. Cherry is a very 'wet' wood, and the raw planks will take some time to become useful timber. Much of the timber that Martin mills remains on site to dry out in his yard, but we decide instead to return our cherry trees to the wood from whence they came, so that we can watch the maturing process at work. In the old, self-sufficient days timber for use at Greys Court would probably have been seasoned in this way well before use. We follow the trailer back to Grim's Dyke Wood, but this time it is laden with slices.

Planks cannot simply be laid out to season. Instead, the log is reconstructed, in its separate slices, but with pegs inserted between each plank to keep them apart, and allow air to circulate. Martin tells me that the tree is now said to be 'in stick'. Poplar wood is employed to make the pegs ('no good for anything else'). The outer part of the sliced tree is just a curved edge covered with what remains of the bark, and this slice remains on top as a kind of umbrella to keep rainwater off the planks below. The overall effect is rather curious, both a decon- struction and an expansion of the original tree trunk. We decide to stack our wood 'in stick' in the Dell, where it is protected from the common view by untidy holly trees, just in case somebody takes a fancy to a plank or two. A few of the best slices go to Philip Koomen, to his workshop in the deepest part of the Chiltern Hills, where he has a kiln that can take out the last few per cent of moisture from the timber before he makes the cabinet to house the collection.

Saved by the chair

As a network of new canals spread across England during the first half of the nineteenth century, bulk transport of goods and raw materials became easier than it had ever been before. The installation of modern pound locks and better towpaths helped Thames trade in equal meas- ure. However, these indubitable improvements could well have spelled the end of the Chiltern woodlands. Demand for firewood in London fell as coal from the Midlands and the North became readily available to supply domestic and industrial purposes. The first lichens to choke on sulphur must have turned a crispy brown colour at this time. The staple 'crop' that had maintained beech woodland for centuries was no longer in demand. Many of the beech trees now had the chance to grow into finer specimens. Recall that John Stuart Mill had noticed 'real woods, not copse, that is, they are not cut down for fire-wood, but allowed to grow into timber, though not to any great age', when he walked through our patch in 1828.

A curious account was published almost half a century later by William Black, a once-popular writer now sunk into an obscurity to rival that of Cecil Roberts. *The Strange Adventures of a Phaeton* is a travelogue dressed up as a novel,[1] and Black, who was an artist, creates word pictures of the various places visited by the eponymous carriage. The book describes the drive from Henley to Oxford as 'one of the finest in England'. The phaeton leaves Henley by 'the Fair Mile, a broad, smooth highway running between Lambridge Wood and No Man's Hill [part of the Fawley estate] and having a grassy common on each side of it'. The writer clearly had a map in one hand and a note-book in the other, and we are left in no doubt about the improvements to the Bix Turnpike since the previous century.[2] Towards Nettlebed the phaeton 'plunges into a spacious forest of beeches', which could be a description of this stretch of the route as it appears today. The woods had changed, and if they could not make money, they may have faced an uncertain future. The fate of the trees in the grand parks like Fawley Court was assured for the time being, but they were stage props in a designed landscape rather than part of a commercial enterprise. The spectre of mattock and grubber loomed over the woods that had survived all previous threats since the Norman Conquest.

Woodlands were saved by carpentry. We know from Daniel Defoe's description and from the Stonor papers that chair-making was already a significant trade in the eighteenth-century Chilterns. During the nineteenth century it burgeoned. Factory assembly of the finished articles hugely increased productivity, and High Wycombe in Buckinghamshire (nine miles north-east of Henley) became the rapidly expanding centre for a new industry.[3] Beech was the standard material for legs and stretchers. Other timbers like ash, elm and oak had their parts to play, but beech was the workhorse of the common-or-garden chair, and beech was drawn in from miles around. Ordinary folk could afford beechwood chairs; they are still common items at country auctions, and sell for next to nothing. Between 1800 and 1860 the number of workshops in High Wycombe grew from about a dozen to 150. By 1875 the output was reported to be 4,700 chairs every day. I cannot help wondering how there could have been enough bottoms to sit on all those chairs. Perhaps people went out in the evening to sit on extra chairs away from home. In 1873 the American evangelists

Dwight L. Moody and Ira D. Sankey proved such an attraction that 19,200 chairs were ordered from High Wycombe to accommodate the posteriors of the faithful or the ready-to-be-persuaded. It is a strange thought that this source of comfort for numerous Methodist behinds originated less than two miles away from Sir Francis Dashwood's diabolical dungeons at West Wycombe. A higher-class and more complex design familiar as the Windsor chair spread around the world, and especially to Moody and Sankey's homeland. These popular seats provided comfortable repose for many an honest member of the bourgeoisie. That they were well-made is proved by the fact that there is one in every antique shop.

A change in the way timber was sold accompanied the furniture boom. It was marketed while still in the woods as standing trees. A plentiful record of these sales survives as advertisements in newspapers, and as printed sale notices. They announce the sale of such items as 'eight thousand scragging faggots'[4] and 'capital butts', which could sound indecent in any other context. Auctions were held at inns close to the woods up for sale, and the lots were 'shown' *in situ* by estate employees to prospective buyers.

The greater part of Lambridge Wood was now owned by the Misses Stapleton of Greys Court, who lived quietly together in the big house for decades. The tithe assessment of 1842 records them as possessing 159 acres of beech woods. Their woodland managers included several generations of the West family, who held the tenancy of Brickfield Farm near the house, and later lived in the old cottage closest to our wood. An auction in 1848, addressed to 'timber merchants, wheelwrights, and others', was shared by the 'Ladies of Greys Court' and Lord Camoys of the Stonor estate.[5] It was held by James Champion & Son at the Bull Inn, Nettlebed, on Tuesday, 15 February, at 'Two o'clock, at which time precisely a Hot Dinner will be provided, at 1s 6d each to be returned to purchasers'. This deceptive largesse was presumably a device to deter time-wasters in search of a free lunch. The Greys Court part of the sale comprised '380 loads of BEECH and ASH averaging about 15 feet in a stick [trunk]'. As a final inducement there were 'good roads and no turnpike to the

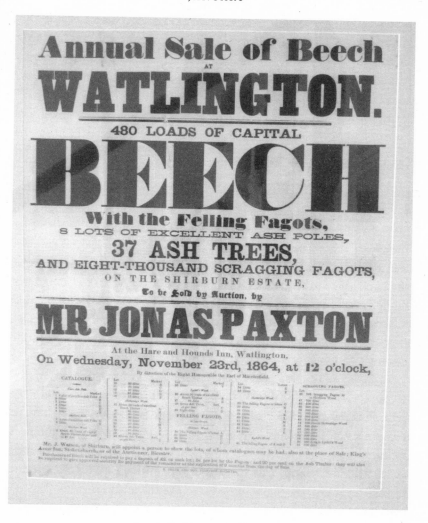

Poster for timber sale, 1864.

Wharfs at Henley' for those who wished to export their purchases up or down river.

Champions had sold two hundred loads from Lambridge Wood nine years earlier at the Catherine Wheel in Henley, so our woods were being regularly worked. Dozens of similar sales throughout the Chiltern Hills and adjacent parts of Oxfordshire ensured regular management of the estates for a steady return in profits. A poster of another auctioneer, Jonas Paxton, shows the Earl of Macclesfield in 1864 selling 480 loads of 'capital beech' from his ancient property at Shirburn a few miles along the Chiltern scarp from Nettlebed. Once again, the endurance of an ancient landscape that dated from Saxon times proved its worth to its stewards a thousand years later.

Bodgers and turners

In Grim's Dyke Wood I nearly fall into a bramble-filled pit on the edge of one of the smaller clearings. Armed with strong gardening gloves I pull out the pesky, thorny interlopers by the roots to discover what they were concealing. Brambles find ways of swiping your exposed bits when you least expect it, and a bloody brow is the result of uncovering this piece of archaeology. It is a price worth paying. The pit had clearly been excavated, as it is encircled by a bank of clay-with-flints – the material dug from the hole. It is about the size of a grave that might have accommodated a particularly large cow; it is a well-preserved sawpit, a legacy of the way trees were formerly processed right here on the ground.[6] After the beeches had been purchased they were felled, and planked by hand. A pit had to be dug to allow a gigantic two-man saw free rein. In 1745 the botanist Peter Kalm described this method of processing 'green' wood by creating a pit 'a fathom deep', 'on each side lined with boards so that earth does not fall down into it'. The same practices continued throughout the nineteenth century, and into the twentieth. I believe mine must be one of the very last examples, because it is still in fine condition and clearly defined. I have seen old sawpits elsewhere in Lambridge Wood which are partly filled in and smudged out after a hundred years of erosion.

The two-man crosscut saw was a fearsome instrument, longer than the average human and with teeth that would not disgrace a great white shark. Examples are preserved in the Museum of Rural Life at the University of Reading, where the 'Do Not Touch' sign must be one of the most superfluous notices in museology. To process a log into boards the trunk was placed lengthwise along the pit, and one sawyer

The 'top dog' above a sawpit at a Herefordshire sawmill in the 1930s.

was straddled above while the other one was downstairs. He was the one who got all the sawdust in his eyes as the saw went back and forth, as well as being imprisoned in a small space. He was known as the 'underdog' – the man aloft in the open air being, naturally, the 'top dog'. So now you know.

The first stages of chair manufacture were carried out in the woods, but have not left such obvious pitfalls. Legs and stretchers were turned on lathes set up under the trees by artisans who came to be known as 'bodgers'. 'Green' beech was again employed for the purpose, and workers made temporary camps in the glades of trees in which they had an interest. The simple shelters they constructed for themselves

'Bodgers', c. 1930: one holds a beetle and wedge, one is using a draw shave, and the third holds a roughly-shaped leg. The hut contains a pole lathe.

were known as 'hovels', and they often camped out in them. There were a few skilled craftsmen who could make a whole chair, but most bodgers were piece-workers who turned out thousands of legs to order. Beech was first roughly cut into lengths appropriate for chair legs ('billets') before being turned.

In the early 1930s Cecil Roberts visited 'the last of the chair makers', a little old man over eighty with a 'bright eye and an impish face', and described the 'crazy old apparatus he called a lathe': 'A sapling ... was bent down, and a cord from this passed around the future chair leg, and was connected to a strip of wood that acted as a treadle. On pressing the treadle down the springy sapling made the cord taut and revolved the chair leg. The turning was done on the down stroke, which revolved the leg towards the worker. Despite the jim-crack nature of these pole lathes, as they were called, a skilled worker could turn out four dozen legs an hour.'[7] The bodgers used the young saplings in the wood to power the processing of the more mature neighbouring trees; no doubt they even used the shavings to boil their billycans. Could a more ecologically sensible industry be devised? The legs were not just tapered spindles, either, but could be considered shapely, often having a raised ring around them purely for ornament.

When they had finished their work the bodgers would take off to an ale-house to spend too much money on beer. The Crooked Billet at Stoke Row was still such an establishment when I first ventured into the Chiltern Hills more than thirty years ago. Tucked away down a side road off a side road, only ale could be purchased, and the barrels sat in full view on cradles in the back room. The bungs were turned beech wood. Oak beams, low ceilings and wide fireplaces attested to the age of the place, which had once been run by a bodger called Bill Saunders. A low murmur of voices provided the only musical accompaniment. Nowadays it is a gastropub.

I am puzzled why the word 'bodge' has come to describe slipshod work. *The Oxford English Dictionary* tells us that it transmuted from 'botch', but this still does not explain why skilled woodworkers acquired such a label. I once had a beechwood chair that fell uncom-

Chair legs and stretchers, produced by bodgers c. 1930,
stacked to dry in the woods.

promisingly into its component parts, but that was a new one, and I have several old chairs that are as reliable now as when they were made. Maybe the bottoms of some of those 19,200 Methodists were let down. The temptation is to romanticise this rural craft, but there is plenty of evidence that it was poorly paid and provided a hard life. Even at the end of the bodging era between the two World Wars the rates were exploitative: a gross of forefeet (144 front legs) and nine dozen stretchers, six shillings (thirty pence today).[8] Bodgers like Bill Saunders did two jobs: running the pub by night, and working on billets delivered to his shed by day. Some of his colleagues were on the wrong side of the law and the woods, as always, provided cover. Others fell into a poverty trap by taking out advances (with interest) against their wages from unscrupulous shopkeepers that they were never able to pay off completely. Medieval Greys Court was probably no harsher.[9] In Buckinghamshire, lacemaking by the women of the household supplemented the family income – a cottage industry so exacting that their eyesight was often compromised. In 1817 Mary Shelley lived with her husband Percy Bysshe Shelley not far away along the Chiltern Hills: 'Marlow was inhabited (I hope it is altered now) by a very poor population. The women are lacemakers, and lose their health by sedentary labour, for which they were very ill paid.' It is indeed altered now: the cottages once occupied by lacemakers are owned by retired army officers and City traders.

Alistair Phillips is a wood turner with a modern lathe. He grew up on the Warburg Reserve at the end of Bix Bottom where his father Nigel was warden, so the Chiltern woods have always been his home. He now lives in the house once occupied by Vera Paul, who devoted her life to ghost orchids and to setting up nature reserves where animals and plants might enjoy a measure of protection. In his workshop Alistair is going to turn a few rounds of cherry wood left from our felling into bowls. The wood is still wet, so cracks could present a problem as it dries.

First, a central strip of the heartwood has to be cut out, as that is where the desiccation cracks generate: it is flanked by a pair of curved planks, in each of which hides a bowl. The two potential bowls in the log nestle side by side like twins in a womb. The bottom of the bowl faces outwards – on a curved surface this maximises the size of the finished article. Like Martin Drew, Alistair is a man of few words and steady concentration. For each plank, a pair of dividers marks out the diameter of the bowl that might be expected. Then, using a bandsaw, a 'blank' with a circular outline is cut out, together with a hole in the centre that allows the bowl to be attached to the chuck of the lathe. At this stage, all the character seems to have disappeared from the wood; it is just an attached block. When the lathe begins to spin fast, water is thrown out in some quantity – we are sprayed with it. The centrifugal force expels it from the vessels of the wood, something that would have happened more gently if it had been allowed to season slowly. Alistair first cuts off the irregularities from the block with his chisel so that it will spin effortlessly. Then he gouges with great assurance, and curls and chips of wood are thrown out like bubbly lava from a spitting Catherine wheel. The turnings are crisps of thin wood, and smell vaguely sweet, the fruitwood smell. He works steadily from edge to centre, employing a concave-bladed wood chisel sharpened to great keenness. Each cut makes a sharp circle moving inwards, until a special twirl finishes off the middle. The base of the bowl is shaped to be very slightly concave, so it won't wobble on any irregularity of the surface it rests upon. An elegant side profile has emerged at this stage. The block is then reversed and the inside of the bowl is brought to life in a similar way. Alistair measures his progress with calipers, because he cannot risk going through the base of the bowl. He wants the edge to be thin, as this ensures that there is less chance of a crack developing later. Already the grain of the wood has reasserted itself as lovely, arched lines give the new bowl character; the warm pink-brown colour so typical of cherry wood emerges. The work is quickly finished off. Alistair tells us that we have to leave the bowl in a cool, dry place to let it season.

I am looking at the bowl now, more than a year after Alistair's turning. It did indeed warp as it dried out, in a pleasant way, so that its rim gently undulates; it has become something of a sculpture. The bowl was partly cut into the sapwood, which is now a deep golden colour, the heartwood darker, but still a pale warm brown on which the growth rings of the tree are painted like contours on a map. They are the contours of a life, an archive of the passing seasons, good and bad. Our cherry bowl is an encapsulation of the recent history of Grim's Dyke Wood, written in the language of the living wood.

Saved by the train

As the writer of *The Strange Adventures of a Phaeton* appreciated, great improvements in the roads made travelling across country a pleasure rather than a test of endurance, particularly when the chances of having fob watches stolen by Isaac Darkin and his colleagues had been reduced. At the end of the eighteenth and early in the nineteenth centuries Henley-on-Thames lay on a main route across the Chiltern Hills, and provided an ideal stopping-off point for travellers to Oxford, and beyond to Cheltenham and Birmingham. The smart and elegant town, now all dressed up in its Georgian finery, was generously endowed with coaching inns, which can easily be recognised today by their wide, low-arched entrances alongside the main premises. Stabling was provided at the rear. Some of these inns had been hostelries since medieval times, like the Red Lion at the end of the bridge over the Thames; rebuilding to accommodate the new travelling classes apparently ensured a profitable future.

These were great times for the Catherine Wheel and the Bull. Services along trunk routes ran through Henley several times a day. *Jackson's Oxford Journal* for 14 February 1824 lists three morning services to London leaving from the Angel Inn, Oxford, including 'The White Horse post coach every morning at 9 o'clock through Henley to the White Horse Fetter Lane, and Angel, St Clements'. In 1838 coaches from Henley in a variety of livery sped in all directions: *Defiance* went to Abingdon and Oxford, a white coach pulled by grey horses; *Tantivy* to Birmingham, decked in red; *Magnet* to Cheltenham, in blue. The cobbles in the Market Place must have rattled to the stamping hooves of dozens of horses. This speeding traffic would have clattered past the wood down the chalky hill towards the Fair Mile carrying passengers from the Midlands. Business and pleasure accelerated to the hoofbeats of the coach and four, and letters, at least for the well-heeled, were delivered with astonishing rapidity.

For a while, business at the wharves continued much as always, but now facilitated by easier passage along the River Thames for the export of the wood and malted barley that had first made the town prosper. It was not to last. By 1840 the Great Western Railway had already reached Reading, and four years later Oxford station opened. Passengers and goods could be whisked by steam engine in an hour or two to destinations that would have taken much of the day to reach by coach. The coaching trade collapsed very quickly. *Tantivy* and *Defiance* were no longer required. The many coaching inns were soon short of custom. Even the Red Lion closed temporarily in 1849. Henley was no longer on the main road.

The wharves were equally in decline. Paintings of Webb's Wharf in 1889 by Janet Cooper in the River and Rowing Museum show the trade in its last days, a slightly decrepit vessel unloading a log or two. *Pigot and Company's Royal National and Commercial Directory and Topography* of 1844 lists the employment in the town. There are only two wharfingers itemised, Isaac Charles and Robert Webb, the same number as there are hatters. Compare this with six leather-workers, nine 'corn and hop dealers and mealmen', ten butchers, sixteen bakers and twenty 'retailers of beer', and it is clear that the town was transformed. The banker Grote left Badgemore House, no longer willing to endure the five-hour journey to London by coach. Only the woods were unaffected by these changes, because the furniture trade operated through a different network, in which the merchants of Henley were irrelevant. The town itself may well have seemed destined to become a backwater on a road to nowhere, a place of former glory, dreaming of past times by its splendid bridge under the beneficent gaze of the Chiltern beech woods.

Something had to be done. Townsfolk, tradesmen and gentry alike surmised that Henley's attractive riverside setting was an asset that could still bring trade, but of a different kind. The Great Western Railway passed through the small town of Twyford on its way to

View of Henley Bridge, with the Angel Inn and St Mary's church,
and a laden barge, 1834.

Reading. A spur off the main line running along a pretty part of the
Thames Valley for five miles from Twyford to Henley would bring
back the visitors. Memories of the formerly fashionable Henley
endured; a day trip might be just the excursion for newly affluent
middle-class families. In 1857 the branch line that still runs today was
opened to the sound of bells ringing in St Mary's church. Lord Camoys
from Stonor House, as always representing the oldest family in the
area, announced publicly the dawn of 'a new era of prosperity' now
that there was 'a means of making numbers acquainted with the
delightful scenery of the neighbourhood'. Here was an important
change in perception: for centuries the woods and dry valleys were
working landscapes (not least at Stonor), a source of billets and bavins,
wheat and wool. The River Thames was as much a waterway as it was
a source of power and of fish for the table. Now a concept of landscape
that had developed on estates like Fawley Court could be applied

more generally. The whole countryside – that remarkable 'living fossil' – was to be appreciated for its beauty and pleasing proportions, by a general public.

That is still how it appears today. For thirty years I was a commuter on the train that runs from Henley to Paddington station. This journey takes longer today than it did at the end of the nineteenth century. The return journey in the evening entails changing at Twyford to the branch line, and on to a small train with three carriages and a chatty clientele. Those who sternly stuck to *The Times* on the main line suddenly smile and remark on the weather, or grimace and talk about the state of the world. As the little train clatters along the Thames Valley through Wargrave and Shiplake a huge sense of relief sweeps over me: I am leaving London for the 'real' countryside. Meadows by the river still look lush. A few pleasure boats tootle downstream as we cross the Thames beyond Wargrave, where willows lean down over the water's edge and a stately grey heron might be spotted standing motionless in the shallows. The wood-clad hills beyond rise sharply in the distance where the river broaches the Chilterns and a few white Georgian houses grace the prospect, as if placed specially to please a landscape artist. Willow and alder carrs, stables and post-and-rail fences protecting fine-looking horses, the occasional mock-Tudor residence – all together induce a species of apparently rustic contentment that make the difficulties of the working day recede into unimportance. Nothing jars; even the trading estate on the outskirts of Henley is discreetly screened. I have fully subscribed to the recreational view of landscape of the fleeing townie.

The recipe for Henley's regeneration worked well. Smart money moved there. A neo-Gothic stately home with 120 rooms, Friar Park, was built on the edge of town close to Badgemore. Sir Frank Crisp (1843–1919), a hugely successful solicitor who lived there from 1889, designed many of its singular features. Crisp was also a microscopist[10] who studied those creatures (he would have called them 'infusoria') too small to be included in my inventory of Grim's Dyke Wood. His house is famous on several counts. Horticulturalists know it for its astounding rockery, built to resemble the Matterhorn. Jackie went to a primary school there run by nuns of the Salesian Sisters of St John Bosco when the building was no longer owner-occupied. She recalls light switches made in the likeness of friars' noses that had to be tweaked every time they were turned on or off. The house was dotted with anti-clerical jokes, which the nuns apparently did not resent at all. There were mysterious grottoes, even underground lakes (out of bounds). The remarkably elaborate house is most illustrious, however, as the home of the Beatle George Harrison, who bought it in 1970, and lovingly restored it. His widow still lives there. Only modern pop aristocracy could afford to maintain such a fantastical monster. I regret that I have never had a chance to see it for myself.

While the Stapleton family lingered on at Greys Court in charge of our wood, there was change at Fawley Court. Although it is a digression, I have to mention Hugh Edwin Strickland, a renowned geologist and naturalist who spent his formative years at Henley Park, more or less opposite Lambridge Wood on the northern side of the Fair Mile. The house was part of the Fawley Park estate, and still stands today; it was originally the dower house for the mansion of Fawley Court by the Thames.[11] During the 1820s, between the ages of eleven and seventeen Hugh learned the rudiments of the natural history that would become his life. He was already a prodigious collector. Strickland Freeman, the nephew of Sambrooke Freeman, who then owned and lived in Fawley Court, was himself an equine anatomist and botanist, and would surely have encouraged the young enthusiast. There is no way to prove that Hugh wandered through Grim's Dyke Wood, but what budding naturalist would not thoroughly explore the woods within a mile of his home territory? He made notes on the snails around Henley that would one day be the subject of a scientific publication.

I cannot help but identify myself at the same age with young Strickland – a time when it is possible to know *everything*: all the flowers, birds, moths, fungi, stones – even the very microbes. Young naturalists are probably a curious sport of nature, like musicians with perfect pitch, and they do conform to type in their essentials. I suppose now they would be called 'nature nerds', but there is something heroic about such enthusiasm to tackle the impossible. There is simply too much diversity in the world for a single head. Strickland had a good try: he published extensively on birds, geology and many other aspects of natural history, and helped set down the 'rules' for scientific naming of animals that we still use today. Another small connection I can't pass over: a fossil brachiopod named for him *Stricklandia* was a contemporary of my 'own' animals, the trilobites. Hugh Edwin Strickland met a tragic end in 1853, killed by a train as he examined a rock section alongside the tracks near Hull. His watch had stopped at twenty-nine minutes past four.

In the same year as the naturalist's untimely death the Fawley estate passed from the Freeman family and their relatives – itself a process not without complications – into the ownership of the Scottish railway entrepreneur Edward Mackenzie, brother of the brilliant civil engineer William Mackenzie. Trains inevitably seem to dominate our story for a while. The new owner was installed in time to see the Great Western Railway transform his new home town just a few years later. The Mackenzies were largely self-made: theirs was new money for new times. Edward's son Colonel William Dalziel Mackenzie inherited the estate in 1880; I have already mentioned his purchase of the land right up to the edge of Lambridge Wood. On the Fair Mile he donated land for the creation of an isolation hospital, financed by W.H. Smith, the stationer, who now owned the Hambleden estate (yet more smart money). Nothing illustrates the difference between the Freemans and the Mackenzies better than their family mausoleums in the graveyard of the old parish church at Fawley. The Freemans' is classically modelled in Jurassic stone topped with a cupola; the Mackenzies' an austere, dark, oblong granite construction engineered to last.

The seal on Henley's revival in the age of steam was the Royal Regatta. It is still the signature of the town. A regatta course is possible because the Thames is unusually straight between Temple Island and the bridge, like a Fair Mile on the water. The first Oxford vs Cambridge boat race was held on this stretch of river in 1829, presumably because it was both suitable and accessible to eights from each university.

Henley Regatta, 1896.

Although the race moved on to London, the regatta tradition stayed in Henley. By 1839, an event comprising several races overlooked by grandstands and with peripheral entertainments attracted a notably fashionable crowd, especially the nobility dutifully recorded in the *Oxford Journal*. Stewards were appointed to oversee the events, as they do today, including – of course – Lord Camoys. Prince Albert was solicited for royal patronage in 1851, and Henley Royal Regatta was established. The railway brought the crowds six years later. The Prince of Wales and the Kings of Denmark and Greece arrived in 1887. Thirty-four thousand people attended over three days in 1895. The coaching days had never seen such a throng. Emily Climenson called it 'the world's water picnic': 'All sorts and conditions of men and women resort to it in thousands, many not caring a straw for the races, but for the charms of the scenery, the kaleidoscopic glow of colour and form represented by every variety of craft.'

It remains much the same today, with as much jollity as serious spectating inside the enclosures set up alongside the course, and many festive small boats pottering about on the far side of the water.

Champagne corks pop, young people fall in the water. Old boys in brilliantly striped blazers greet one another to celebrate survival for another season. Nonetheless, Henley hosted Olympic events in 1908 and 1948, and has been a famous place for international crews to test their relative strengths ever since. The Leander Club on the far side of the bridge is the leading rowing organisation just about anywhere, and home to Britain's best competitors: it boasts more Olympic medals than any other club in the world. The only problem is that there are no longer enough local hotels since the nineteenth-century 'slump'. Henley residents host crews for the Royal Regatta, so for one week our house is full of muscular giants from Cambridge University. Pasta consumption quadruples to supply slow-burning carbs.

The meaning of the landscape around Henley was changing. The annual Regatta shindig had no interest in woods as a productive resource, but rather as one component of an attractive setting. Train services to London meant a new class of commuters could work in the city and return at the end of the day to comfortable villas on the south side of town. Sir Frank Crisp could receive his clients at home. The connection to London that had been important since before medieval times now had nothing to do with exchange of goods, but rather with the direct transfer of money and people. The Civil Parish of Henley was redrawn to accommodate new housing and eliminate part of the ancient 'strip' of Rotherfield Greys that ran down from the hills to ensure river frontage in feudal times. Old affinities were irrevocably severed. Not for much longer were the many ancient paths that crossed the landscape the arteries of daily employment, although they were still there to be reinvented for recreation. Our wood was set to acquire a new role – but not before some dramatic interruptions.

Snow

Overnight, several inches of snow have settled in the wood. A slow, steady fall of big flakes has left every holly leaf with a burden of white icing. The tiered branches of the small yews, usually so discreet and dark, are suddenly blatantly arrayed for a winter festival. The tops of even the smallest horizontal beech twigs have been decorated with a snowy crust; it is as if they have been created anew as black-and-white. Beech trunks have snow plastered on one side, making *ad hoc* cascades, or shuffled ripples or pressed-on beards. The boles of some trees wear white trousers, and they peel off slightly at the top, as if rolled over.

It is supremely quiet. Where lumps of snow have tumbled from above on to the blanketed wood floor beneath they have made one of those white-on-white abstract paintings that hang in Tate Modern. Our footprints are the first human ones, but others have gone before us. A hare (could it be the same one?) has lolloped through the wood making pairs of flat ski-prints with its long back limbs, and dints ahead of them.[12] Neat lines of paw-marks must be those of a fox after wood mice. In the time I have owned the wood I have never set eyes on Mr Fox himself; his town cousins are far less cautious. A scuffle under the King Tree has cleared away the snow and revealed the leaves – is this evidence of his search for prey? I have heard mice at work under the nearby holly trees in the autumn. Split-toed muntjac deer prints track across the clearing – in search of any green shoot, I imagine. All these shy creatures are more easily seen from the tracks they have left than met face to face. On the snowy wood floor we mammals are all joined together in the democracy of our foot-prints: pad, paw, hoof and wellington boot, each diagnostic of our species.

'Argh!' Jackie cries, pointing. 'There it is! The orange stuff!'

She gestures towards an ash trunk. One side of it appears to have been painted bright orange, a shade rendered almost garish in contrast to today's snowy background. There is more of it elsewhere in our wood, but particularly in the patch of Lambridge Wood adjacent to our own, where it gaudily decorates beech trunks over several yards. 'I don't like it … there is something sinister about it,' Jackie mutters. 'I'm sure there's more of it than there used to be.' The patches are clearly organic, so my first thought is that they must be produced by some kind of lichen. Now that we have been made aware of the orange thing, whatever it is, it does seem to be common. Hardly a country stroll is possible without my wife protesting about it taking over. 'See!' she cries, as if it is somehow my fault. 'There it is again.' Microscopic examination proves that it comprises yellow algal threads. It is a species called *Trentipohlia abietina*, common as one of the partners in the lichen symbiotic association, but perfectly happy to live on its own. As so often in nature, the yellow pigment is a carotenoid, and probably protects the alga from solar radiation, which is why it can live in such open positions. I shall have to add a small piece of this stained bark to the collection. I cannot find a particular reason to find it more sinister than, say, the green alga *Desmococcus olivaceum*, which I often find as a bright, dusty-looking crust on the shady side of the boles of our beech trees. That species looks as if somebody spilled some of the old-fashioned green powder paint we used in school.

Trentipohlia is assuredly more noticeable. My wife is not alone in her conviction. Several naturalists have gone online to point out a similar phenomenon in their patch; it is extraordinary how the Internet allows observations to be shared so quickly. The orange paint may well be spreading. My algal friends are not sure why. It may be that *Trentipohlia* – like some lichens – is becoming more common as sulphur air pollution declines, or possibly it is encouraged by those unseen sources of nitrogen which are proving bad for fungi. I dread the thought of all our smooth and shining beech trunks becoming robed in unsubtle orange.

As we leave the wood a small rodent, probably a bank vole, scuttles desperately along the road in front of us before apparently diving into a snowdrift. It is the only living creature we have seen today. My wife says it is probably running away from the orange invasion.

11

February

The moss man cometh

Beech trees and holly bushes are suddenly full of small birds. Long-tailed tits signal to one another with sibilant whistles, blue tits chatter wheezily, and I spot finches somewhere in the mix as well. They all flit restlessly from branch to branch, bobbing acrobatically, possibly in pursuit of hardy midges that have appeared with warmer temperatures. Hanging together must increase their chances of foraging success. The ambient light is as bright as it could possibly be, and although there is no sign of any fattening buds, an intangible message is passing through the trees that winter will not last forever. Lengthening days signal to dormant plants, stirring them into action. Teams of fat wood pigeons work together through damp leaf litter under the trees. For once, they are living up to their name; the prevalence of these birds on my own vegetable patch suggests they should have been dubbed 'garden pigeons'. They have an air of plump self-satisfaction about them – a puffed-up busyness – that makes me want to take them down a peg or two, and not solely in revenge for my broccoli plants. Their white ring of neck feathers like a clerical collar increases my irritation, giving them a holier-than-thou gloss. Yet when I walk towards them they rise as one bird and flap away with short wingbeats to a part of Lambridge Wood where they will not be further disturbed. They are as smart as they look. I wonder what they seek: possibly a few beechnuts missed in the autumn feeding frenzy. On the fields beyond the wood, grey fieldfares all the way from

Scandinavia peck at tiny seeds to help see them through the toughest time of the year. In this season it does not pay to be too fussy.

Winter branches, Grim's Dyke Wood.

Much of the wood still sleeps through the winter, but for some of its inhabitants the top season of the year is right now. Ridges of earth thrown up next to ruts, larger flints lying on the edge of Grim's Dyke, bare soil, boles of trees and many of the higher parts of the trunks are decked out with green patches in a dozen subtly different shades. Feathery yellow-green whiskers climb up the stems of ash saplings. Crisp little shoots run over the surfaces of the old pine woodpile, and their tips shine almost silvery in this light. A round cushion of deep green offers a tempting seat by the wayside, but I know from experience that actually sitting on it will soak my backside. Mosses thrive everywhere, rejoicing in the abundant winter light freely given to them while the beech trees are dormant. In hot summer they are often no more than crispy patches, easily overlooked. But now they cannot be ignored, and it is obvious that there are many different species. In my naïve enthusiasm for the Grim's Dyke project I thought I could identify them for myself, and requested as my February birthday present a thick handbook of all the British species, published by the British Bryological Society. I soon found myself floundering; there are a *lot* of different mosses, and most of their differences are subtle. I need help.

Peter Creed knows all about mosses and liverworts. As befits his specialism, he is a gentle soul. He moves through the wood with that special kind of intense concentration I have seen in so many naturalists. I imagine the same stance adopted by Aboriginal trackers in the Australian outback, able to spot the signs that a goanna had passed by two days previously. Liverworts are little more than creeping, branching, photosynthesising pads; many botanists believe that something rather like a liverwort made the first colonisation of land from the waters more than four hundred million years ago. I am used to seeing relatively flamboyant, crimped species growing along damp banks, but Peter points them out to me in my own patch, where I had missed them completely. Creeping over damp surfaces of the rotting pine pile, a green film resolves under my lens into flat shoots a few millimetres wide with 'leaves' on either side – I am told this is crestwort (*Lophocolea heterophylla*).

I can forgive myself for missing this pretty little species, but I have a harder time to pardon my failure to spot liverworts on the trunks of beech trees. Peter finds one tree on the edge of the clearing with lots of pale-green patches closely pressed on the trunk – my hand lens reveals most of them to be made of flat, almost seaweed-like, forking strips, the yellow-green thalli of forked veilwort (*Metzgeria furcata*). A flatter-looking scalewort (*Radula complanata*) is growing on the same tree, like a series of tiny overlapping green tiles. Then, higher on the tree, what looks hardly more than a cobweb is the smallest plant I have ever seen, with shoots a few millimetres long, but with round 'leaves' on either side of the midrib which must be about a quarter of a millimetre in diameter, like a string of minuscule peridots. If Peter had not pointed out this plant to me, I would have passed it by forever. 'Fairy beads,' he tells me. '*Microlejeunea ulicina*. Generally more typical of the wetter western parts of Britain.' So that makes three species of liverwort on one tree, and I had missed them all! The surprise is that they seem to be able to grow even where there are no mosses, as trailblazers on tree trunks. I had thought that liverworts were inhabitants of damp corners. They are more adaptable than I realised.

We move on to look at old tree stumps, which are richly decorated with enough moss to turn them into upholstered stools. Rough-stalked feather moss (*Brachythecium rutabulum*) is the commonest decorator here, thickly covering the surface almost like shag-pile, with pointy tips of the growing shoots much paler, bordering on silver. This species is mixed in with the commonest moss in the wood, which, thanks to the guidance of my kindly moss man, I now observe growing just about everywhere, even climbing along twigs near the ground: common feather moss (*Kindbergia praelonga*) has shoots that resemble miniature, delicate fern fronds. Peter tells me to check the differing shapes of the minute leaves on stem and branches with my hand lens. Some of the patches of this moss bear capsules that carry the spores; they rise on slender red stalks from the sides of the shoots, looking very much like elegant swans' heads, complete with bill. A third moss is almost as common, cypress-leaved plait-moss (*Hypnum cupressiforme*), which does resemble tiny matted shoots of cypress foliage, as its Latin name helpfully suggests.

I am starting to get my eye in on the differences between various kinds of moss as Peter pads on through the wood, peering at a likely tree or picking up flints. 'They are all acid or neutral soil species,' he advises, which squares with my own geological thoughts. On the bank, the tempting dark-green 'cushion' is a neat, close mat of erect, unbranched shoots, each bearing relatively long, slender, lance-like leaves arranged in rosettes – the bank haircap (*Polytrichastrum formosum*) is a variety that people instinctively stroke as they pass a mossy bank. A somewhat similar, but much paler moss on the same bank has wider leaves, with beautifully crimped edges (common smoothcap, *Atrichum undulatum*); its capsules have long beaks shaped like golf tees, and they rise like miniature charmed cobras from the green carpet below. A fork moss (*Dicranum*) grows with it, bearing leaves thin as hairs all curving in one direction. Different mosses evidently grow in separate places in the wood, some preferring well-drained ground, others happily colonising wood or stones. I remember the sawpit, surely the dampest spot in the wood, and when Peter climbs down into it, sure enough he discovers three kinds of moss we have not found before, including juicy silk moss (*Plagiothecium succulentum*) and thyme moss (*Plagiomnium rostratum*), the latter with large (well, for a moss) rounded leaves so thin as to be transparent. Near the dyke, on mossy stones (ones that have never been rolling, which as we know, gather no moss), another silk moss (*Pseudotaxiphyllum elegans*) sports curious feathery branchlets extending off its dangling shoots.

Mosses are easy to store for the collection; they are just dried for an hour or two and then popped into small packets. The number of species in the wood climbs to fifteen, but there is nothing rare enough to have escaped inclusion in Peter's guide to local species[1] until we find a small moss forming neat, almost spherical cushions on fallen branches of cherry and beech. We had earlier found a greener, straggly version of this moss growing on beech trunks, which Peter had identified as the wood bristle moss (*Orthotrichium affine*), another common species. This one is different, with compact little red-tipped capsules projecting from its surface like diminutive gooseberries. Our strongest hand lens reveals a few long hairs at the base of the little beak (seta) that tops the capsule. '*Orthotrichium stramineum*,' Peter announces with confidence. 'Straw bristle moss. Not in my book. Nice find.' This small moss has only been found in this part of the world a couple of times before. It is common in Wales. Peter thinks it has found the air quality more to its liking recently, although it is only thirty miles or so from the middle of London to Lambridge Wood. The abolition of coal fires in the city in the last century may have begun to pay dividends at last in ways only a 'moss man' would instantly appreciate. Grim's Dyke Wood can smell the change in the air.

Beeches: the last stand

The wood is inured to change; it has been accommodating changes for a millennium. At the end of the nineteenth century it was still working for a living. The most striking change was in the sounds that reached it from the road to Bix. Since turnpikes were abolished in 1888, steam-driven traction engines had taken over much of the work of dragging logs and performing heavy duties on the farm. The puffing, wheezing and gasping of the engines and the crunch of their huge, metal-shod wheels on the newly surfaced road must have made the whole area feel more industrial than rural at busy times of the year.[2] Granite road surfaces were added in 1909 to help cope with the wear. The hills were short of water, and traction engines needed a ready supply. Close to the old tollhouse at Bix at the top of the hill a large, rectangular tanked pond still remains to show exactly where the engines slaked their thirst. For a decade or so in the twentieth century horses, traction engines and early internal combustion engines co-existed in the Chiltern Hills, before modern petrol-driven tractors became universal. Every summer, steam-lovers still charge up their restored traction engines to congregate at Stoke Row near the Maharajah's Well. They seem appropriately elephantine. All their bits and bobs of brass are polished to golden perfection, and all their paintwork glows. One engine is deployed to power a steam organ, to jolly effect. The work that goes on is for display purposes only. As for horses, they might still provide the least damaging way of dragging out a large, felled log from the wood, if only I could find the right kind of powerful beast to oblige.

Edwardian Henley was a town of elegant indulgence and leisured luxury, and if it was built on the back of an Empire that stretched its capitalistic tendrils over half the world, I do not suppose the average spectator at the Henley Regatta gave that a second thought. The house I live in was built on the junction between the reigns of Queen Victoria and Edward VII, and is a comfortable villa made for the middle classes. There are bell-pushes in most of the rooms to summon the maid. They no longer work. A gentleman who was 'something in the City' might have commuted to and fro, leaving his family to enjoy a healthy country life, with weekends together punting and picnicking on the Thames. Mine is an appropriate house for a geologist, as it has limestone surrounds to the windows and panels of local flint for decoration.

View of the centre of Henley in the early 1900s, along the side of the new town hall towards Hart Street.

The original inhabitants would have been able to enjoy their privileged lifestyle for a dozen years before everything changed. The First World War awakened Britain from its Imperial complacency – and also transformed the use of the common beech tree. Huge numbers of wooden items were required as part of the war effort. The beech woods provided raw material for rifle butts. Field camps used tents, and tents demanded pegs – many millions of them. What else should provide this necessary, but unacknowledged, essential for life under canvas but the reliable beech tree? Tent pegs were supplied in several sizes. Making a good tent peg requires twenty-four different movements, which were carried out on a primitive wooden sawhorse. In its way, peg-making was as complicated as turning out a chair leg, although the cause was more urgent. Like many skilled artisans, 'peggers' had their own special tools, with names that I find strangely poetic. Beech logs fresh from the woods were split into appropriate lengths using a huge, odd-shaped wooden mallet called a 'molly' that engaged with an iron splitter known as a 'flammer'.[3] The workers sheltered under corrugated-iron sheets held up by four poles, with draped sacking to keep out the draughts. Wives and children helped with the stacking of the finished articles.

Many of those who might have been servants in more leisurely times were called to the trenches, or became workers for the war effort, women included (this may have been when my bell-pushes fell out of use). The 'peggers' were not above exercising their industrial strength, even in wartime. The *Reading Mercury* of 22 April 1916 reported: 'About 20 tent-peg makers employed by Mr. Douglas Vaderstegen … struck work at Stoke Row on Monday morning. It appears that the men, who were recently granted an increase in the price of large size pegs, asked for a rise of 3d per 100 on small pegs.' Whatever the success of their action, their pay was still niggardly. The demand for pegs continued during the Second World War – in 1942 the Stallwood family at Stoke Row received an order for no fewer than two million pegs. I recall the same kind of tent pegs being used on camping trips during my childhood, whacked in with a wooden mallet, but their production on an industrial scale has now ceased for several decades. It has pegged out.

The nearest sawmill to the wood was at Middle Assendon, along the valley to Stonor. To reach Froud's, an Edwardian traveller from Henley would have turned right on to the smaller byway at the end of the Fair Mile, where a pub called the Traveller's Rest Inn occupied the fork in the road. This hostelry had to be demolished in the mid-1930s when a second carriageway was added up the hill on the main road to Nettlebed and Oxford – which became one of the first 'dual carriageway' highways in England. Froud's sawmill in Assendon operated for a century from 1866. It was 'green'. Waste wood products were burned to power its large stationary steam engine, with a forty-foot chimney in the yard surrounded by stacks of sawn timber seasoning slowly, and unsawn trunks piled up awaiting attention. Martin Drew would have felt at home there. Froud's was by no means a small operation.

At its height between the wars it employed nearly a hundred men, who walked to work from all the surrounding villages along ancient rights of way – today's public footpaths. The mill was known for making the wooden backs of brushes. It had a national reputation among brewers for its turned shives and spiles (two more wonderful words!). Shives were the bungs in wooden beer barrels, and made from hazel – a reason to maintain some woods with the ancient 'coppice with standards' system. Spiles were the vent pegs in the barrels, usually made of beech. Froud's was bustling when Cecil Roberts moved into Pilgrim Cottage in Lower Assendon. In one of his thinly fictionalised novels based on the area he described the passage on a 'chariot' past his cottage of the beautiful young son – nicknamed Apollo – of 'Farmer Lowfoot', 'with fine hair fluttering above his brown head and throat. In the early sunshine he passed like the young god driving up the dawn. There followed the blower down at the wood yard calling the men to resume work.' The manufacture of spiles and shives seems a mite ordinary after raising the sun.

Beech from our wood was destined to go a little further afield than Froud's mill. After nearly nine hundred years, the manor of Greys Court was separated from Lambridge Wood when the last Stapleton owner sold our ancient woodland to Mr Shorland in July 1922. An era of quicker profit was upon us. By 1938 Shorland had sold on his interest to the Star Brush Company. Their works at Stoke Row were a major local employer, and vast quantities of beechwood were fed into the factory from woods all around to make the backs of brushes. I have the price catalogue for the company for 1925 ('Patent machine-made solid back brushes'), which lists a great variety of scrubbing brushes, laundry brushes, nail brushes, hat brushes, clothes brushes, shoe brushes, stove brushes, horse brushes, and more. Prizes were awarded to the company at international trade fairs in Paris (1878), Sydney (1879) and Melbourne (1880), for 'merit and novelty'. Rather wonderfully, a handwritten note to a customer that fell out of my catalogue added (in pencil): 'We do not make brooms.' One of their products might linger in the back of a cupboard in your house: a sensible board carrying a heavy set of bristles and a longer scrubbing set at the front.

Indestructible. The bristles listed in the catalogue are of 'Mexican fibre', the product of the tough agave shrub. For paintbrushes, the best bristles came from pigs raised at Chungking in China.

I feel certain that fine beech trees from Lambridge Wood found their way into the backs of countless humble scrubbing brushes, which were items of even less common regard than wooden spoons – the traditional bottom prize in any competition. How are the mighty fallen! The trees used for brush manufacture were usually more than seventy years old, and at the height of the business 100,000 cubic feet of beechwood were used up every week. Contemporary photographs record huge machines relentlessly gobbling up the timber. Nevertheless, the Star Brush Company was proud enough of its reliable products to put its six-rayed brown star trademark on the back of every item. Only obliging beech could tolerate the rigours of the milling process. Star had factory premises in Holloway in north London, where the finished products were assembled. The company made profits, too. Although the depressed interwar years were tough (net profits 1938–39: £1,860), the shortage of imports during World War II boosted the sales of the home-made products enormously, such that company profits for 1944–45 were £25,842.[4] In 1955 the Star Brush Company merged with Hamilton Acorn, a long-established Norfolk brush-manufacturer. Increased mechanisation and 'economies of scale' led inevitably to such mergers as the British economy struggled to recover from the war years.

Brush-back making in the Chiltern Hills continued until the Stoke Row premises finally closed in 1982. Cheap imports and the invention of even cheaper plastic products finally ended the beech tree's last stand. By that time our wood had already been in the ownership of Charles Darwin's descendant Sir Thomas Barlow for thirteen years, and it was no longer expected to yield any kind of regular 'crop'. Many trees were free to grow to maturity to help develop the woodland dog-walkers and joggers enjoy today; and, of course, to provide shelter for the occasional scavenging naturalist.

Very few old-timers who worked in the Chiltern beech woods survive to describe the daily lives of the woodsmen. David Rose is one of them. Now in his eighties, he lives with his wife Mary in an old cottage in the valley on the edge of the Stonor estate at Pishill. Mary worked as Lady Camoys' personal secretary, while David worked in the woods, so they connect viscerally to the Chiltern landscape and its embedded human history. They are both full of life, and welcome a chance to talk about the old days. David tells me that his grandfather was a gypsy who married on Salisbury Plain, then settled in Barn Cottage at Maidensgrove, uphill from the Assendon Valley. His father was one of twelve. David's aunts wore long black skirts, remained spinsters, and spoiled their young nephew. His rubicund, gently folded face loses its ingrained good humour momentarily as he recalls, 'My father was hard as iron.' David went to school with Maurice MacRory, who later ran the sawmill near Nettlebed which is still operated by the same family in 2015. David started work at fifteen. Some years later, in 1955, he cut a hundred acres with his father 'with axes and crosscuts', which went to Froud's for processing. The family lived in the Market Place in Henley, travelling to where their skills were needed, 'even as far as Windsor Great Park'. David was employed as a piece-worker all his life: 'No dole for me; sometimes I only earned ten quid a week. Sometimes there were big jobs that paid well. In 1968 I cut 100,000 cubic feet for the Andover Timber Company. I worked alongside horses until specially designed tractors replaced them.'

The harvesting of timber was sustainable; small trees were protected from damage to replace the felled generation. It was not always so. In 1934 Cecil Roberts lamented mass wreckage: 'a terrible and heartless devastation, and slopes once glorious with ash and beech stand riven and naked to the sky'. David Rose supervised the felling in Lambridge Wood when Sir Thomas Barlow harvested some sections (not ours) on the advice of Mr Mooney. With permission, wood could still be cut there even though it had been declared a Site of Special Scientific Interest. He recalls felling a huge ash – 'a beauty' – in Lambridge in 1968, but he had to be careful: 'It tends to split and kill the feller.' 'Some of the lads who helped in the woods were no better than they should be: "hobos" we called 'em. In the summer they lived out in a pig shelter.'

David's was a tough life. He recalls hearing the drone of Lancaster bombers over the woods during the last World War, sometimes 'with smoke coming out'. On 29 May 1942 a Spitfire came down near the Fair Mile, killing a Polish airman. The crash would have been heard in Grim's Dyke Wood. Even the Chiltern Hills could not escape the turbulence dominating Europe. Missions took off from a local A-shaped airfield at Kingwood Common, no more than four miles west of our wood: 'All completely overgrown now, though the concrete runways are still there, under the trees.' I have explored this former airbase; it somehow seems to belong to a deeper archaeology than the twentieth century. Fawley Court was requisitioned by the British Army for training special forces, and many commandos passed from its peaceful riverside meadows into the thick of the fray in France. After the war, the former seat of the Whitelockes, Freemans and Mackenzies was purchased by the Congregation of Marian Fathers, for use as a school to educate Polish boys displaced by the tragic events in their home country. 'Mind you,' said David Rose, 'I never felt in as much danger during the war as I did in the great drought of '73 to '74. There were lots of "stags' heads" [dead branches] through the woods, and "widow-makers" were falling all the time.' He smiles gently at the thought, an old man at ease with his memories after a long life well spent in the company of beech trees: the last of his kind.

Last orders

I find evidence of the woodsmen who worked in Grim's Dyke Wood. Half-buried in the beech litter a couple of bottles attract my attention. My first thought is: Litter louts! Just jettisoning their bottles when they'd finished with them! One of the bottles is brown, with a design etched upon it and a heavy screw-in stopper still in place; the other is green, with a black screw stopper with a milled edge. Traces of a rubber seal linger on its underside. My first impression was wrong: these are old bottles. They were probably meant to be re-used; their glass is thick, and they are altogether heavier than their modern counterparts. Matching the threads of the stopper with the inside of the glass neck took careful manufacture. The etched design on the brown bottle is rather elaborate: around the perimeter of an elliptical cartouche the words 'Brakspear & Sons Ltd. Henley-on-Thames', and in the centre a nicely rendered honeybee underlain by a scroll labelled 'The Brewery'. The screw stopper of the green bottle also carries the Brakspear identification as a relief. So there is no doubt that these were bottles of beer produced by Henley's own brewery, which still uses the bee as its logo. They will have to go on their side to fit into the collection. But when were they left behind? This design is extinct. Metal crown caps are almost universal on beer bottles today, although more elaborate contraptions are used to seal some up-market German brands. I have early memories of screw-stoppered bottles of luridly coloured 'pop', but I am fairly certain that by the time I was allowed to drink beer those bottles had already disappeared.

My investigation takes me to an antique shop in Henley. The Tudor House is one of those emporia stuffed with bric-à-brac and china, and all manner of curiosities crammed in from floor to ceiling. It is a place in which I have to keep my arms under careful control lest I nudge something to self-destruction ('Breakages must be paid for'). Tucked around one corner is a shelf carrying old bottles, including one or two like mine. The proprietor of the Tudor House, David Potter, is able to confirm my hunch. He believes that stoppered bottles were used by Brakspear until the mid-1950s, and recalls that there was a deposit of a penny or two on the bottles to encourage people to return them. Now they might be worth £20. So they must have been abandoned at least sixty years ago. At that time Lambridge Wood was still in the hands of the Star Brush Company, and access was not as free as it is now. In the middle of Grim's Dyke Wood it is more than probable that the thirsty people were employed by the company itself – they must have been woodsmen. The growth-ring counts I have made of my trees suggest that many beeches are about eighty years old. Sixty years ago they would have been young trees, of the right age to replace those that had just been felled for processing. David Rose would surely have approved the succession. This was the last serious harvest of mature trees in Grim's Dyke Wood. I envisage two sinewy, red-faced men, resting after their labours. Maybe they were responsible for digging our well-preserved sawpit. The older man takes the more expensive brown ale in the engraved bottle, while his younger colleague has the 'ordinary'. Perhaps he has already done his turn as the underdog, but he does not complain. They both realise that they will never come to work this wood again in their lifetimes. They carefully replace the stoppers and leave the bottles behind to mark the occasion.

I am happy to discover a direct link between our wood and the Brakspear brewery. For well over two hundred years it was as much a part of Henley as St Mary's church, and located right next to it in the heart of town, where a brewery really *ought* to be. The smell on 'brew' days was as distinctive a signature of the municipality as the famous Henley Bridge, and provided atmosphere in the most literal sense. Brakspear (rhymes with Shakespeare) pubs still dot the countryside around the home town, although there are far fewer than there were twenty years ago.

The founder of the brewery in Georgian times was Robert Brakspear, born in 1750, who proved to be a man typical of the age, with a lively interest in rational experimentation to improve his product.[5] He had thermometers and gravimeters to help standardise

This view has changed little in two hundred years, but Brakspear's beer is no longer brewed in the heart of Henley.

several varieties of ale, and he kept detailed and meticulous notes of all his 'brews'. The Henley brewery was never very large compared with those in London. It produced about six thousand barrels a year at the beginning of the nineteenth century. The town had long been known for the production of malt, when barley is sprouted and then quickly dried to release natural sugars. Several of the old maltings survive to this day in different guises. There was good water in abundance, and a ready market. The brewery prospered. When I first moved to the town the old brewery yard was still operating, and I went on a tour of the brewhouse to watch their famous copper still in action and to attempt to understand the fermentation process. The rise in property values made this complex of lovely old red-brick buildings more valuable as 'real estate' than factory. In 2002 they were sold off, and the still was moved to Witney, in the middle of the county, for Brakspear's ale to be made under licence by the bigger firm of Marston's. It felt as if some vital organ had been ripped out of the living town, and I have been unable to pass the old brass sign for the Brewery Office without a shudder ever since. The buildings are still largely untouched externally – they have been partly converted into a hotel and restaurant – but they have lost the purpose for which they were built. The same thing has happened to pubs in the hills that have been converted to private houses. One by one, the old inns have disappeared. I mean the kind of places where a few old boys sat chortling in a corner, and a real wood fire blazed when it was chilly. There is nothing more forlorn than a dead hostelry; nothing more final than when the cry of 'Last orders' means exactly what it says.

Blow, winds, and crack your cheeks!

On a few occasions, worse destruction is wrought by nature. In the first days of February 1990 bewildered foresters were still totting up the casualties of the great wind of 25–26 January that has come to be known as the 'Burns Day storm'. In the Chiltern Hills the hurricane was more ruinous than the famous storm of October 1987 that brought down many of the specimen trees in the Royal Botanical

Gardens at Kew, and flattened large tracts of the forested Weald. The 1990 storm also accounted for more human fatalities, because it happened during daylight hours. Beech trees were particularly vulnerable to sudden gusts of wind that exceeded 100 mph. In the high Chilterns the woodlands were not only exposed, but the shallow chalk soils failed to provide secure anchorage for large stands of trees. Beech roots do not dig deep, but fan out around the base of the tree. The storm was so severe that many other kinds of trees could not escape damage. *The Times* described the devastation in Stonor Park that greeted a horrified Lord Camoys: 'In front of him all 15 ash trees were laid out neatly like stalks of asparagus on a plate. At one end of the house a cypress had toppled and missed the fourteenth-century chapel by perhaps a foot. At the other end a cedar so big, so fat in its girth, that it was hard to believe it was a living thing had been flipped over as if by a wanton finger and had dumped itself on the wall of the shrubbery. Beeches the size of factory chimneys lay prone everywhere, dozens and dozens of them.' The flanks of the Assendon Valley fared particularly badly: whole beech woods were laid low. The toppling of one big tree set off an arboreal domino effect, with one tree after another wrenched out of the ground only to collapse on to further trees and spread devastation deeper into the forest; felling on a scale that exceeded any ever carried out by human hands. The aftermath was a no-man's-land of tipped trunks and matted and broken branches.

A quarter of a century later, the sites of woodlands that were destroyed in 1990 are still obvious as areas of scrub on the flanks of the hills. They have been replanted, but it will take another fifty years for new trees completely to erase the memory of the Burns Day storm. The tipped-up root plates of the fallen beech trees are easily spotted: lumps of white chalk dragged out of the ground by roots mark out the bases of the dead trees, and the holes they once occupied have not yet filled in.

In another sense, the storm revived the wood. Long-buried seeds germinated; breezes and birds brought in other plant species. In the recovering woods near Stonor are masses of thistles and brambles, St John's wort, and chalk specialists like gromwell and ploughman's spikenard. To a goldfinch or a whitethroat this is a much more congenial home than a high beech stand. Butterflies and moths might find more food plants. More insects would feed more insectivores. It could be argued that short-term devastation is long-term conservation. In the natural order of things there will be unusual storms from time to time, and their aftermath may refresh parts of the ecosystem that would otherwise disappear. 'It's an ill wind that blows nobody any good,' as the old saw says.

Two miles south of our wood, Hunt's Wood is a small piece of ancient woodland that was badly mauled during the 1990 storm. A few enormous beeches survived and now stand alone, their lofty trunks soaring strangely naked in the absence of their former fellows. Lucius Cary is replanting his wood, but with a mix of trees: more wild cherries, whitebeam and oaks as well as replacement beeches. In the end the richness of the woodland may well be enhanced, but it will be Lucius's grandchildren who will notice the difference. Time has healed Stonor Park. A few beech trees still stand along the public footpath through the grounds, making small groves. They are among the oldest trees in the area. Because they were grown as ornamental landscape trees they have branches low down, and are altogether stouter than the woodland stands of straight-trunked, lofty beeches. They have both more gravity and more gravitas. They resisted the hurricane. In 1803 the famous landscape architect Humphrey Repton remarked that most Chiltern beech woods 'are evidently considered rather as objects of profit than of picturesque beauty'. He might have regarded their tumbling down like ninepins as aesthetic justice.

Lambridge Wood (and Grim's Dyke Wood) escaped lightly from the great winds, as Mr Mooney reported to Sir Thomas Barlow. How this happened is difficult to deduce, though I surmise that the narrow Assendon Valley might have funnelled the winds into lethal gusts that particularly affected tree-covered chalky slopes. Our woods are on nearly level ground, and the clay-with-flints soil might have been more stable. Two beeches in our small patch *have* blown over: their root plates still stand vertically, showing how the roots once extended more or less horizontally over the woodland floor. I suppose they remain in place because it would have been too much trouble to remove them after the trunk timber had been cut up and processed. I cannot say whether these trees were tipped over in 1987 or 1990. I can say that they did not set up a chain reaction of falling neighbours. One of them created a small clearing when it fell, and now several small beeches and a wych elm compete for light in the open space. The end of one tree provides opportunities for the next generation. A wood in which no tree ever tumbled would stifle its own future.

I make a quick visit to the wood to ponder and wander, which has become a form of occupational therapy. It is still cold, and a wan mist that Turner would have appreciated filters the distant sun. The new year's bluebells are already spreading pert rosettes of splayed dark-green leaves, and the wintry light has awoken other plants. Tongues of cuckoo-pint leaves (*Arum maculatum*) emerge from the deepest litter, unfurling into triangles, some blotched with black. They are in a hurry to grab energy while they can, to produce starch that they will then store in white tubers deep underground. By high summer there will be no sign of the leaves; only green sticks will remain, rising straight from the ground carrying poisonous berries that ripen red in the autumn, when they seem to have nothing at all to do with the luscious leaves in front of me now. As I scuffle through the old beech litter a robin redbreast pops out from a holly tree and pounces on something edible that I have disturbed, just inches away from me. He too is curious. He is adapted to investigate the ground stirred up by any passing animal; he might think of me as a rather inadequate wild boar. Was it the same robin who had followed the Chiltern Society? I clearly pose no threat; he darts back into the holly cover to see if I will uncover any further treats.

Bird names can be confusing: I remember being shown a 'robin' in the eastern USA, and realising it was a completely different bird from the European 'robin'. An American visiting Europe experiences the converse. The two birds are even placed in different avian families, though they were formerly both lumped together with the thrushes. Nowadays, ornithologists have popped our cheery little redbreast into another evolutionary packet altogether (an old-world flycatcher, no less), while the US version is still related to thrushes. Maybe using Latin names would help, but *Erithacus rubecula* somehow fails to convey the quality of 'robin-ness'. I shall stick with the common tag.

Charcoal

An elderly correspondent writing in the *Henley Standard* in April 1922 recalled charcoal burners at work in Lambridge Wood in the latter years of the nineteenth century. Clearly this is something to try. Weight for weight, charcoal is a much more efficient source of energy than firewood, and commands a premium. It was exported to London from the wharves at Henley for centuries; in the Forest of Dean and the Weald it provided the high heat necessary for smelting iron long before coal took its place. The principle of charcoal manufacture is simple enough: to drive off the sap and other volatiles from hardwood to leave behind pure, black carbon. A fire creates heat that is then sealed beneath the wood to be combusted, expelling all the fluids and impurities. Manufacture of charcoal in large quantities required a very skilled hand to ensure proper stacking of the wood, and to build a covering of soil or turfs to seal in the combustion. The pile had to be continuously guarded for several days to prevent 'burn-outs' – an exhausting business. Fortunately for me, charcoal can be made in small quantities using nothing more than an old oil drum. Andrew Hawkins is my guide to charcoal manufacture at the level of a cottage industry.

One end of the oil drum is cut out to improvise a lid; the other has a smaller-diameter hole cut into it to act as a chimney. Any old dry twigs and small boughs lying on the woodland floor can be used to start a lively fire. Then the oil drum is placed over it to provide a shelter for the blaze. Now the hard work begins. A stock of slender fallen beech branches (some squirrel-damaged) has been gathered together, and the job is to saw them on a sawhorse into lengths of about eighteen inches, and throw them through the chimney end into the drum on top of the fire. Boughs of a width about the same as my wrist are ideal. Sawing is hard work – it does not take long to build up a sweat. I can imagine the hard labour that must have gone into stacking a full load. The edge of the base of the drum is very slightly levered up on one side to provide just enough air to the fire. Once the drum is half-full, prolific white smoke emerges from the 'cooking pot' – to avoid choking it is as well to check the wind direction before setting up the sawhorse. White smoke is the sign that the wood is in process of transformation, but unlike the election of a new Pope, this white smoke is the beginning rather than the end of the process. At this stage the lid is loosely propped atop the chimney exit to keep the heat sealed in. Now all we have to do is wait for an hour or two, until the colour of the emerging smoke takes on a bluish cast. This is the signal that the wood is becoming 'cooked'. At this point, all air is excluded. Loose soil is piled around the base of the drum to seal it, while the lid is weighed down over the chimney hole with a heavy stone. The edges of the lid are sealed tight using a sand fillet around the perimeter (sand falls back into any vacant space as it dries). Now we can go away and leave the sealed hot drum to its own devices.

The following day we return to unpack the booty. There is a decent return of nice blocky charcoal – less than a third of the volume of the wood that went in. A coarse sieve is used to take out ash resulting from the odd piece of wood that has gone just too far. Andrew tells me that before he got the method right he once had the experience of finding nothing but ash for his pains. Now we have two large bags of excellent natural charcoal to fuel our barbecues. As commercial charcoal is often made from mangrove wood, damaging a major tropical habitat and a valuable shield against coastal erosion (to say nothing of tsunamis), we can also take home a sense of having done the right, sustainable thing.

12

March

An unexpected discovery

The vernal equinox approaches. A clear dawn among the trees induces a sense of anticipation that infuses the very air. Lambridge Wood has woken from its winter sleep already, but the beech leaves are wrapped within their buds awaiting some arcane signal to instruct them to unfurl; the sun can still reach everything under the trees. Light briefly catches on the wings of insects on the move, and twinkles on dancing gnats. Scuffling noises like scrunching paper from under the leaf litter prove that small mammals are secretively scurrying about their business. Large, grey-spotted leopard slugs have emerged from hiding. As for the birds, mating has become imperative. The whole wood resounds to the drumming on hollow trees of red-and-pied greater spotted woodpeckers. They are advertising themselves with a rapidfire tat-a-tat; a snare drum rather than a kettledrum, but a cleaner noise than either, and designed to carry. There are two – maybe three – in different parts of the woods. The repetitive percussion is only possible because they have built-in shock absorbers in their muscles, and a hinge between skull and beak. Human headbangers sustain more damage, and probably attract fewer mates. So there is an orchestra in the wood today: a wind band of birdsong – chaffinches with their descending cadences, rich flutes of blackbirds, chimes of great tits – with the woodpeckers adding their percussion section *allegro con brio*. Then, quite suddenly, a discord intrudes on the symphony: the cackle, the mad hilarity of a green woodpecker, the cry my father called a 'yaffle'. For a few seconds a penetrating series of repetitive, rather harsh notes dominates the woodland. It is like a whoopee cushion interrupting a concert. I spot the culprit escaping heavily through the trees, a large green bird flying, I should say lolloping, in almost dolphin-like fashion towards the open fields beyond the wood. The symphony soon resumes with undiminished vigour.

Around the clearing a few offcuts from the felled cherry remain to be carried home for splitting into logs, so there is work to be done. Laura Henderson, a friend who works for the Forestry Commission, is being introduced to the wood. Brambles grow too vigorously in the open. They originate from gnarled woody bases that are very tough. I had always assumed that briar pipes must have been manufactured from this wood, but I discovered that the raw material is actually derived from a Mediterranean tree heath (*Erica arborea*) for which the local name is *bruyere*,[1] so tobacco pipes will not, after all, be one of my country products. *Ceci n'est pas une pipe.* Long, arching branches splay out from the old bases, and root in their turn, soon making an impenetrable mesh of prickly vines, ideally designed for tripping up people carrying logs. Just now, shoots for new branches crouch like rockets, ready to go when the starting pistol fires.

Laura hollers excitedly from the far side of the patch, near a pile of twigs left behind from the felled cherry. She has found a perfectly round, woven nest the size of a tennis ball, with a circular entrance hole in one side of it. The nest has been constructed of long grass leaves – wood melick, I would guess – that have been perfectly wrapped in a number of layers like a hollow ball of wool. A few dark strands might have been cherry bark. It looks like a cosy place in which to snuggle up. Enlightenment comes in a few moments. It is a dormouse nest! This is the first evidence in our wood of one of Britain's most delightful mammals, and one that can be described as 'cute' without blushing. It is also a protected species under the Biodiversity Action Plan, so the discovery is important for the conservation of a rare species. It is hard to stop jumping up and down with excitement at the discovery that there are dormice in Grim's Dyke Wood, but the nest seems so fragile and easy to damage, and bramble tripwires lurk everywhere. I might drop the evidence. Gingerly, the precious object is borne back to the vehicle, as if it were the golden hoard. We cannot be certain exactly where the nest originated: it could have been tucked among the brambles, or possibly it came from high in the felled cherry. It will be a special item for the collection.

One of the advantages of being cute and uncommon is that people take a special interest, something that does not often happen to rare woodlice. Jackie describes the big-eyed, orange-brown, fluffy-tailed dormouse as 'wiffly-piffly', which may not be a familiar word, but sums up their attraction concisely. The National Dormouse Monitoring Programme (NDMP) is partly run by the People's Trust for Endangered Species (PTES), so acronyms are on our side. A photograph of our nest emailed to a helpful scientist was quickly confirmed as showing a dormouse hibernating structure, which is 'not uncommonly' found in bramble patches. Our record will be added to the national database. Dormice are well-known from the Warburg Reserve[2] a couple of miles to the north of the wood, where they are protected and encouraged to breed, so it is not so astonishing to find them in Lambridge Wood as well.

Dormice (*Muscardinus avellanarius*) are famous for sleepiness above all else, as they 'spend up to half the year in hibernation and are torpid for much of the remaining time'.[3] The behaviour of the Dormouse at the Mad Hatter's Tea Party in Lewis Carroll's *Alice in Wonderland* was not so much fantasy as scientific description. The ability to slow down their metabolism is an important survival strategy. When they are awake, flowers, fruits and nuts are important foodstuffs; the little mammals need to pack nourishment away to see them through their protracted sleepy periods when food is less available. Recent research shows that dormice live high in the canopy at some times, and close to the wood floor at others. Grim's Dyke Wood offers bramble flowers, hazelnuts and beechnuts, but there is no honeysuckle, which is one of their sugary staples (though I have seen this plant elsewhere in Lambridge Wood). Like fairies, they dine on ephemeral delicacies. I have to wonder what they seek out in springtime in Grim's Dyke Wood. We have no hawthorn flowers for them to sample. Perhaps they wake up early enough to enjoy the wild cherry blossoms that abound in the canopy in April, though this does not figure in research studies I have discovered. No matter: I just welcome these charming animals into our wood, even if they manage to suck nourishment from dew and dreams.

Man's estate

The story of Grim's Dyke Wood has proved inseparable from the histories of the surrounding estates, and more particularly the manor of Greys Court, whose lords owned the ground for so many hundreds of years. Attitudes to rural property through the twentieth century changed in unprecedented ways – although, curiously, the large estates have come full circle. The greatest transformation concerned the humblest dwellings. Small cottages once occupied by blacksmiths, or coopers, or chair-makers, are now the proud possessions of businessmen or professionals or the comfortably retired. A solicitor will attend to his electronic mail in a room that was once foggy with carpenter's sawdust. Just next to our wood, 'the murder cottage' and the barn next to it were always part of the same estate as Lambridge Wood, and they were sold off together in 1922. They are part of our story. The cottage is now a charming country residence, tastefully extended in an appropriate fashion, surrounded by an ample and lovingly tended garden. Next door, the former barn has a huge picture window in its side, and has been converted into a comfortable family home.[4] A newer house has been erected at the back of the older buildings on the site of the old orchard. When Cecil Roberts lived at the bottom of the hill in Lower Assendon in the 1930s he regarded the houses at the top of the hill as out-of-the-way, and at the time of the Dungey killing in 1896 the *Henley Standard* emphasised the unusual remoteness of the spot. The modern internal combustion engine changed all that. Across the Chiltern Hills historic cottages have become valuable assets. Up at Crocker End, on the old route to Nettlebed, I visited a 'cottage' that may originally have belonged to one of the brickmakers. Whole wings had been added, constructed of brick-and-flint, a landscaped garden surrounded everything, and two racing cars snuggled down on the drive. If it had not been done so accurately and 'in keeping' there might have been something offensive about this vernacular ennoblement. The hills are scattered with pretty villages, and many of them have no local people left in them at all. The village of Turville, lying three miles north of the wood, is known to millions from an aerial view at the beginning of the hugely successful television comedy series *The Vicar of Dibley*. It presents as remotely rural. In fact, it is almost entirely occupied by people who work in the media. The cottages looking so tickety-boo would once have had vegetable plots in their front gardens, and maybe even a pole-lathe in the shed.

Cecil Roberts lived in Lower Assendon at a time of transition. He admired the skilled, but ageing workers in the village, and mopped up their stories from his station in the Golden Ball; he even provided financial help to several local people in trouble. But he had *Gone Rustic* – to cite the title of the first of his Pilgrim Cottage books in 1934. His choice of title was significant. Some of his friends in the village were decidedly genteel, like Miss Whissitt, who conversed in idiosyncratic French. The 'county' set was still more or less in place. But Cecil was committing to something of an experiment in a strange land – a rustic adventure (with added housekeeper and gardener). The flavour of his books is reminiscent of the popular 1989 work by Peter Mayle set in rural France, *A Year in Provence*, in which humorous episodes with 'real' *paysan* people are combined with tempting glimpses of a more authentic life. It proved a heady mixture. Many Britons have headed for the south of France on the back of it. I am not saying that Cecil Roberts was in any way responsible for changing the demography of the Chiltern Hills, but he reflected a change in *Weltanschauung* that became ever more pervasive. By the time H.J. Massingham had published *Chiltern Country* in 1940, 'gentrification' was well advanced, much to that opinionated author's horror. Villages were changing, and the meaning of rural life was changing with it, with the woods as silent monitors.

Many ancient manors were under threat during the twentieth century. Estate duties on larger properties were introduced in 1914 by the Asquith government, and as these taxes gradually increased they took their toll on stately piles when they passed from one generation to the next. Income from Empire declined as the great British trading nation was embroiled in two world wars, and several stately homes – like Fawley Court – were requisitioned for military purposes; some of them never recovered. 'New money' took over. Lambridge Wood had been sold in 1922, and in the following decade Miles Stapleton disposed of Greys Court and its associated farmland, ending the long association of his family with the parish of Rotherfield Greys. Cecil Roberts volunteered to mediate on behalf of several of his friends interested in purchasing it, although the manor buildings had by then declined to a sorry state. In 1935 the remainder of the Greys Court estate was sold to Evelyn Fleming, who restored the house with more *élan* than wisdom. She already owned a large house at Nettlebed ('Joyce Grove') on the back of banking wealth. Her sons were the travel writer Peter Fleming, and Ian Fleming, creator of James Bond, which makes for a briefly glamorous association, but two years later she had sold on to the Brunner family.

Their money derived from Sir John Brunner, a brilliant industrial chemist and co-founder of Imperial Chemical Industries in 1926. His son Sir Felix (1897–1982) restored Greys Court much as it remains today, converting parts of the ancient fortifications into delightful walled gardens, including the finest and largest wisteria pergola I have ever seen. As a Liberal Member of Parliament he proved to be a politician enlightened enough to introduce statutory sick pay for ordinary workers, and he engaged with the local area in a way that Henley had not seen since the days of Sir Bulstrode Whitelocke. His wife Elizabeth was equally public-spirited, becoming chair of the Women's Institute National Federation and vigorously promoting adult education for women. Lady Brunner lived on in Greys Court until her death in 2003, at just short of a hundred years of age. Hugo Brunner followed his parents' tradition of public service as Lord Lieutenant of Oxfordshire. They were, to use the black-and-white classification of

1066 and All That, a Good Thing. Felix was also a great supporter of Octavia Hill, founder of the National Trust, and in 1969 Greys Court was given to the Trust to secure its future. The house has currently been 'frozen in time' by the Trust exactly as it was when it was the Brunner family home, and I don't doubt that they deserve their starring role, although we have seen that their tenure was a grace note on an extraordinarily long history, of which Lambridge Wood was a part. I cannot but wonder what the reclusive Stapleton sisters would have made of all the National Trust visitors gawping at their greenhouses.

Our other manors followed utterly different trajectories. *Nobody* can see them. Badgemore House, which owned some of the lower stretches of Lambridge Wood, passed into the hands of another prominent local family, the Oveys, in the later nineteenth century. Richard Ovey still farms the Hernes estate, by the road from Henley to Greys, and he showed me photographs of an elegant, finely proportioned house, with well-dressed members of his family posing in front of the pedimented entrance designed by Christopher Wren's master carpenter Richard Jennings. By the 1930s the Oveys had the now familiar problem of maintaining such a large house, which was not in a good state after the First World War, and they sold it on to a man called Vlasto, who built a new and more convenient house immediately to the north. The stable blocks were hived off as riding stables for a Miss McAlpine. Both houses were requisitioned in the Second World War, and the older house was demolished in 1946 – after a fire, according to Richard Ovey – and vanishes from our story. The newer house and the stables became the nucleus of the Badgemore Golf Club, which still operates today.

More big houses around Henley survived Hitler, and might well have been destined to follow Badgemore House into perdition. Instead, new employment as educational institutions rescued them from destruction. Friar Park, Fawley Court and Park Place all became schools after the Second World War; for youngsters, evacuees from Poland, and children with problems, respectively. At this point the history of the big houses takes a bizarre turn: the estates have gone back into private hands. Friar Park was bought first, by George

Harrison, and thus passed into the domain of a special kind of royalty. Very wealthy people bought the other houses. Fawley Court is currently being refurbished for a reclusive private owner, with 'no expense spared', after the Polish school closed and the Marian Fathers finally departed. Park Place was the most expensive private house sale in English history at £140 million, sold in 2011 to a Russian oligarch called Andrey Borodin. I know nothing about him, and it is not wise to enquire too closely. All I will say is that he has refused to allow a bat survey of a special site on his grounds which has been catalogued every year for the preceding twenty; I do not suppose we would have much in common.

Almost every remaining estate around Henley has been purchased by a Swiss banker, Urs Schwarzenbach, including the parkland on the other side of the Fair Mile from our wood. He has acquired much of the Hambleden estate to the north that was once the property of W.H. Smith, of stationery fame. He added Culham Court – a splendid small stately home three miles from Henley – as his private residence, and topped it off with Henley Park, in Fawley parish, combining his estate holdings under the inelegant, if accurate, label Culden Faw Estate. He converted much of his land to service his polo team, the Black Bears. Polo is a sport that can suck up infinite quantities of money, so it is a convenient indulgence for the very wealthiest. Most of the fields around Henley are now fenced with double post-and-rail, and are as neat and as lacking in biodiversity as a croquet lawn.

So far as I know, none of these hugely rich people ever emerge from their secure gates to come into town. I did hear that the wife of Mr Borodin visits a neighbour by helicopter, but that is just gossip. They might as well be on private islands, which I presume is exactly the idea. This new aristocracy does not have deep roots in the location, though I imagine (I am obliged to imagine) they appreciate the woods as they might an expensive painting on a far wall. On the other hand, I must recognise that this is history recycled, in a perverse way. Colonel Mackenzie, Strickland Freeman and Bulstrode Whitelocke all added to their holdings of land, and the right kind of marriage brought the Knollys and the Stapletons additional estates. Today's new aristocracy is an international caste, and their takeover could be viewed as the final stage of engagement of this corner of England with the rest of the world that started with a young Knollys buccaneering in Elizabethan times. What is different now is the exclusiveness of the landowners: they really *do* exclude.[5] Just about everybody, unless they arrive in a helicopter.

Beetles

Beetles are the most diverse group of animals on earth. Their front wings are modified into tough covers (elytra) that conceal a pair of flying wings that can be deployed when needed, so this group of insects can get down into crevices and under the ground – or indeed almost anywhere. Their larvae can make a meal of almost anything. So beetles occupy an astonishing range of ecological niches.

Ground-living beetles are different from those that live in standing dead wood, or on flowers or dung, and many species are nocturnal. During the course of the Grim's Dyke project I attempted to sample some of their different habitats. This took much of the year, so the beetles appear here in the last chapter, although I have already mentioned one or two discoveries that were made in the canopy with the help of a cherry-picker in June. A trap for flying beetles was a special piece of 'kit' that was suspended from a branch well above the ground; it was equipped with a lure and a series of cones that helped any beetles to tumble into a sample jar. A different trap for ground-dwelling species was easily made. I half-filled a jam jar with diluted Dettol (rather like Pernod, it goes cloudy when water is added), to which I added a couple of drops of washing-up liquid, which breaks the surface tension so that anything that falls in is submerged quickly rather than struggling for hours. I buried the jar so that its lip was level with, or very slightly below, the surrounding ground. After a few days the contents were tipped into a preserving jar primed with alcohol, for later examination.

I was astonished at just how many beetles must have been march-ing around the wood after dark – during summer, the jar was fairly crammed with them. The disadvantage of this collecting method is

that slugs also fall into the trap, and when they do a mass of unsa-
voury slime befouls the mixture. I fished out the slugs with a spoon
before proceeding further. Common finds in the jar were big, dark
ground beetles, which are terrestrial hunters, and often flightless. The
jar was also stuffed with predatory rove beetles (Staphylinidae) that
are mostly long and wiggly and look rather like earwigs to the unini-
tiated. In fact they have very small wing cases, which allows the back
end of the abdomen to be seen much more readily. The largest species
is commonly known as the devil's coach horse. When I left a sample
jar a little too long in the ground something began to decay, and the
smell probably attracted a distinctive orange-and-black burying
beetle (*Nicrophorus vespilloides*). This beetle is an arthropod undertaker. It buries the corpse of some small animal and then lays its eggs
upon it; the eggs develop into carnivorous larvae that consume the
body. This particular species shows a surprising degree of parental
care, culling its larvae until there are just enough to prosper on their
grisly repast. I am fairly confident that I can identify a shiny, chunky,
black woodland dor beetle (*Anoplotrupes stercorosus*) with spiky legs,
one of the dung-eaters – there is plenty of deer dung in the wood on
which its larvae can feed.

That is about as far as I can go. Many more beetles lurk in the jars
to be identified. Some of them are tiny, just a few millimetres long. I
need the help of a specialist coleopterist or two. In Britain there are
over a thousand species of rove beetles alone, more than four hundred
weevils, and who knows how many wood-borers, fungus-eaters,
pollen-lovers and so on; it began to seem an impossible task to name
my samples. Colleagues from the Natural History Museum came to
the rescue. Max Barclay totted up fifty species or so from the collecting jars, and from earlier collections made by entomologists swishing
their nets. A small brown beetle, *Ernoporicus fagi*, got him excited – it
is another species from the wood classified as 'nationally scarce', with
just a handful of previous records. In company with a young enthusiast called Jordan Rainey, the Museum's beetle curator Michael Geiser
came to the wood in June and snuffled about, spotting beetles just
about everywhere. Michael is short-sighted, so when he caught some

tiny beetle, he simply lifted his glasses and squinted at it to get a magnified view. Then he shuffled instantly through a vast mental encyclopedia of Latin names before pronouncing. I followed behind him with my mouth hanging open. Not all of the beetles could be identified in the field. Some had to be mounted back in the laboratory for careful examination under a microscope. However, it was clear by the end of the day that more than a hundred different kinds of beetles lived together in the wood. Michael thought the real total might be double that.[6]

Some beetles were not a surprise. I had noticed the perfectly round exit holes of 'woodworm' beetles in the standing dead trees, dusted with wood powder, and three small species with this life habit were duly identified. There were several other kinds of wood-borers. The rest of the list made up a recap of many of the organisms we have met in this book – beetle larvae or adults fed on the lot of them. Half a dozen or more fungus-eaters were recovered, some quite conspicuous, like *Mycetophagus quadripustulatus* ('four-spotted fungus-eater') with four red spots (unsurprisingly), but others were tiny, dark and anonymous. One species (*Enicmus testaceus*) specialises in slime moulds. There was even a snail-eater (*Silpha atrata*), with ridges on its black back, that injects its victim to make the mollusc's flesh more digestible. Several diminutive shiny beetles relied on eating pollen. I expected to find specialists for feeding on beech leaves or bark among the list, and I was not disappointed. Further specialists dined on grasses or nettles, many with larvae gorging on roots. Weevils of various complexions were capable of making a meal of just about any seed or plant. Five different click beetles shared the ability to escape predators by leaping up unexpectedly with a loud report. Max and Michael added additional dung beetles and carrion-feeders. It seems that any conceivable ecological trade has an appropriate beetle or six to carry it out. As for predators, these included slim soldier beetles and brilliant scarlet cardinal beetles; while the carabid ground beetles that scuttle away suddenly when disturbed under a log are already up to ten species, and the rove beetles more than a dozen … and the full list is not yet with me.

It was a relief for me to be able to identify at least one big beetle on brambles in July, whose larvae grow slowly, munching deep inside deciduous wood. Huge, knobbled antennae told me it was a longhorn beetle, and striking yellow and black livery narrowed it down to *Rutpela maculata* – the harlequin longhorn. Not every beetle demands ten years of study before it can be named.

The future of woods

Grim's Dyke Wood is a survivor, but its survival until now has depended on being useful. It would not have endured as 'semi-natural' ancient woodland without having a continuing role within the rural economy: for game, fuel, charcoal, chair legs, brush-backs, faggots, spiles – there was always something new to stimulate another generation of tree management, and defer clearance. Under the protective umbrella of the Greys Court estate the woods possessed a kind of historical inertia, whereby tradition might sometimes outweigh expediency. The persistence of the woods is the arboreal equivalent of the snowclone 'The King is dead! Long live the King!' (Perhaps 'The wood is felled! Long live the wood!') The assumption is continuity. This is not a sentimental indulgence, more a visceral feeling that woodland will be safeguarded, that what has already been will continue onwards into the next generation of trees.

But if the market no longer wants timber, where do we go? Beech is unloved. Nobody wants to eat off beechen platters. Very few customers prefer to have their kitchen work-surfaces made of beechwood, though sentimentalists might want to continue to sit on their old beech chairs. Nobody needs to use bavins or faggots, or even beech brush-backs. Up till now, utility controlled the fate of beech forests, and the market dictated how many trees should survive, and to what age; even in 1966 beech was still in demand for furniture.[7] The high priest of woodland history, Oliver Rackham, has told us that there is less management of our tree resources now than at any time in their history. As this is written the chief economic use for beech is to supply chunky logs for open fires. Have beech woods otherwise become little more than rural decoration? They provide a backdrop for pleasurable excursions, part of the landscape through which cyclists pass on their mountain bikes. A generalised *frisson* of well-being is induced through walking or cycling along one of the old tracks through Lambridge Wood; this has nothing to do with sustainable harvesting of beech trees, and everything to do with escape from the workaday world. I wonder if our wood has become a facility, like a public rose garden or a municipal car park.

Consider a comparable transformation in the status of the River Thames. It has changed from the flowing artery that sustained Henley as a trading post for centuries into a plaything for boat people. The pleasure principle was already in charge in 1889 when Jerome K. Jerome published *Three Men in a Boat* – which remains one of the funniest books in the English language. The skiff carrying the three heroes and the dog Montmorency on their adventures passed Henley without remarking anything industrial. Ratty and Mole in *The Wind in the Willows* by Kenneth Grahame (1908) are drenched in the pleasures of the watery life: as Ratty says, 'Believe me, my young friend, there is *nothing* – absolutely nothing – half so much worth doing as simply messing about in boats.'[8] A similar pottering spirit imbues the charming, discursive journeys Robert Gibbings recounted in *Sweet Thames Run Softly*, published in 1940. Gibbings was fully informed about scientific research on animals and plants he encountered along the way, and described their natural history with eloquent accuracy; if something of his eclectic spirit has rubbed off on this book I would be gratified.

Through all these books the Thames flows as a moving spirit, and fosters a beguiling notion that life on the water is leisurely and fun, but also somehow more authentic than, for example, life in the average office. Readers still love these evocations of boating happiness; they engage with them at some deep level. When vintage boats come out of storage in Henley at Regatta their polished wooden elegance and slow pace immediately awaken nostalgia for the days of Ratty and Mr Toad. Their owners are more likely to stress the hard work involved in keeping them in trim and afloat. The modern 'gin palaces' (as my father-in-law called them) are something else: too fast, too white, too large and too opulent. They seem out of scale for the river on which they cruise. They have names like *Maybellene* or *Georgie Girl*, though I did see one called *Conspicuous Consumption* that I can forgive for chutzpah. No matter: these vessels still follow the course of the River Thames for pleasure, and in that regard their owners are not so different from Jerome K. Jerome and his friends.

The market has driven lack of demand for English timber. It is cheaper to import American cherry than to use the native trees – the fine specimens I felled have yet to find a buyer for the planks they yielded. Cheap 'teak' furniture floods the stores, to the dreadful detriment of tropical rainforests. Native English oak maintains its value in 'prestige' furniture and floors, although American oak is a competitor in the wider marketplace. An ironic twist is that for a long time British global reach helped to underwrite the Greys estate as trade and colonial ambition expanded; now, the very prevalence of such globalisation serves to make the wood unprofitable. As the fortunes of the Star Brush Company prove, even war was good for the demand for beech. It's the untrammelled free market that compromises its worth.

Our trees may now be left to age in peace, but that will not be best for the wood as a whole. 'Crop rotation' and selection of trees for felling in the sustainable way that happened over past centuries is better for biological richness. If the wood were allowed to age to a kind of senility it would benefit only wood-eating beetles. The wood needs to be managed; young trees need to replace the old; new light needs to flood in. Fortunately for our part of the Chiltern Hills landscape, much of the surviving woodland is on estates where shooting is still important. The new 'lords of the manor' are wealthy enough to keep large areas for game, even if their timber is not productive. Other patches endure to the north, around Amersham (Betjeman's 'Metroland'), where woodland is a civic amenity in communities affluent enough to 'buy in' management regimes. Then there is the Woodland Trust, a conservation charity that does wonderful work purchasing and managing woods across the UK for the enjoyment of all visitors, with an eye on species that need protection. Harpsden Wood – another ancient beech wood just south of Henley – is but one among more than a thousand woods run by the Trust. Despite these encouraging signs, we cannot know whether these initiatives will be enough to allow our ancient woods to endure for another century.

Maybe the answer lies in a revival of interest in natural wood products. A magazine for people like us is called *Smallwoods*, and the front cover shows men more rugged than I am happily wielding chainsaws or making poles. Hurdles, walking sticks, beanpoles, wood fencing, chipping, charcoal, baskets, wood-turning and wood-carving are all in the purview of these small wood owners. My first walking stick is just the start! Or maybe it is best to reach out to children. I like the 'One Oak Project' sponsored by the Sylva Foundation,[9] which is in some ways like this book scaled down to the size of a single tree. After a 222-year-old oak tree was felled on 20 January 2010, its wood was used for everything from sculpture to sawdust, and hundreds of children watched what was happening. The oaken sawdust went to smoke something delicious at Le Manoir aux Quat'Saisons, the nearest three-star Michelin restaurant to our wood, run by the chef Raymond Blanc. I too have considered offering him some of my sawdust in exchange for his food.

Trees have been hard-wired into our way of thinking since the days when our forebears believed in dryads. Trees afford the best way of visualising descent: in literate societies tree images are universal in linking present generations with their past, in validating claims for lineage.[10] Medieval trees illustrating armorial relationships are decked with coats of arms much like an orange bush laden with fruit. It is not just that branches leading from a central trunk provide a convenient metaphor for common ancestry, but the longevity of trees is appropriate for evoking history, and for the entitlement that comes with inheritance. Notable families are depicted like oak trees in their ramifications and their endurance; I have mentioned special roles for ash, oak and yew in European culture. Early scientists first began to classify animals and plants into groups, and Linnaeus later gave the world a system for naming them. When the idea of organic evolution and modification by descent was added into biology in the 1830s – how else to portray relationships than by employing the image of a tree? Ernst Haeckel's famous 1874 'Tree of Descent', leading from lowly animals to humankind, is quite clearly an ancient oak bedecked with all our animal relatives – although humans are placed at the apex, much as the incumbent Lord of the Manor might top a medieval family tree. Even when modern science (a field known as cladistics) criticised this sort of over-literal portrayal of descent, the notion of trees as displaying closeness of relationships was not abandoned. Indeed, trees with many different topologies now persist into the age of molecular biology.

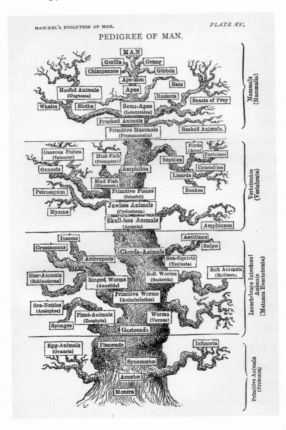

Haeckel's 'Tree of Descent', 1874.

Our emotional engagement with trees is not a matter of genealogy, however; it seems to be more fundamental than that. We find trees reassuring, just as artists find them beautiful. I have spent all this book resisting the temptation to quote the mawkish poem by Joyce Kilmer, 'I think that I shall never see/A poem lovely as a tree,' and it creeps in here with its tail between its legs; but we know what he means. A better, and equally well-known, poem, 'Leisure', by W.H. Davies, seems to hit the mark more precisely:

> What is this life if, full of care,
> We have no time to stand and stare.
>
> No time to stand beneath the boughs
> And stare as long as sheep or cows ...

This brings to mind a jogger who came whizzing through the wood in a pinkish tracksuit one day. She was wearing a set of earphones, and as she went past I could hear a faint 'tish-tish' noise from whatever she was listening to. The songbirds did not stand a chance. She wore dark glasses, and I have the feeling that if she could have run with her eyes closed she might have welcomed the opportunity. I had to fight an urge to stop her gently and invite her to 'Stand and stare! Stand and stare!' But then, I have been too well-brought-up to do that kind of thing.

An apology to all small creatures

That aurally insulated jogger would certainly have failed to notice tiny flies or parasitic wasps. This is where I have to make my own apology. I must offer my regrets that I cannot do justice to all the smallest animals in the wood. They each have their biographies, and there is no logical reason why any one should not have a story to tell as interesting as that of the bluebell or the red kite. There are just too many species of small insects, and some of them have not yet been identified. What I am obliged to do is pick out a few species from an ever-expanding list for special attention if they help promote under-standing of the ecology of the wood as a whole.

I shall start with *Ophion obscuratus*, identified for me by Gavin Broad at the Natural History Museum. It is a special kind of wasp. The insect order to which it belongs (Hymenoptera) is a huge group, including ants, bees, sawflies and vast numbers of tiny insects, as well as the familiar stinging wasps and hornets, all of which share a 'wasp waist' – an attenuated junction of the abdomen with the thorax – as well as several other features on their wings. *Ophion* is an ichneumonoid wasp that we found flying in March. At a couple of centimetres long it is relatively conspicuous as these insects go; its transparent wings carry a few prominent veins, and one marginal cell is characteristically dark-coloured. It lacks the long, egg-laying ovipositor at the rear end so typical of some of its relatives. This species is the first parasitoid I have mentioned by name. Its life habit is a grisly one: it lays its eggs on the caterpillars of moths belonging to a group called noctuids, seeking them out in the dark; one egg per caterpillar. When the larva of the wasp hatches it consumes the caterpillar from the inside, but slowly, in order to allow its host to go on growing until the wasp larva is ready to administer the *coup de grâce*. Then the killer pupates, having successfully turned moth protein into wasp. This parasitoid lifestyle may sound very specialised, but there are thousands of species with this natural history in Britain alone; in fact they are the most biologically diverse hymenopterans.

Most parasitoid wasps are much tinier than *Ophion* – just a few millimetres long. Andrew Polaszek collected plenty of them when he swished a fine net through the herbage along our track. Some of them went off to the Hope Professor of Entomology at Oxford University, Charles Godfray – they were 'his subfamilies'. He kindly identified twelve different species. I was interested when he told me, 'The *Dinotrema/Aspilota* group are currently undoable,' because that is scientists' code for 'We need to do more research before we can recog-nise just how many species there are.' There could be new science to be prosecuted in the wood, and a true expert knows never to impart false certainties.

They may be hard to identify, but parasitoids are hugely important in keeping small pests – like greenfly – under control. They often prey on only one species, to which their physiology is peculiarly adapted; hence they are of unrivalled importance in the biological control of troublesome insects. Research is under way right now to find the 'right' parasitoid to control a moth that is making a meal of horse-chestnut leaves, turning them a horrid brown long before they normally fall. Recent molecular investigations have proved that wasp parasitoids long ago recruited the help of a virus to fool the immune system of their prey caterpillars so that their grubs are able to grow unchallenged. Peculiar and deadly warfare is being prosecuted among the leaves and herbs.

'Bug' is a general term for anything with jointed legs belonging to the phylum Arthropoda. My own particular animals, the extinct trilobites, have been referred to as 'dinobugs' (I don't approve). True bugs are just one group of insects – the order Hemiptera, to which I owe another apology. This group, and particularly inconspicuous sap-sucking 'greenflies', will not get the attention they deserve. I shall mention one bug, the froghopper (*Philaenus spumarius*), which is much more conspicuous because in spring its nymph (the young growth stage) hides itself in a palace of white bubbles on nettle stems near the woodpile. 'Cuckoo spit' is the common name for it, I suppose because its appearance coincides with the arrival in Britain of the eponymous migratory birds – another parasite, of course, which is sadly now becoming rare. When its protective bubbles are smeared away, a green, helpless nymph is revealed inside, vaguely struggling.

As for the true flies (Diptera), the group includes hoverflies, scuttle flies, flesh flies, bluebottles, ensign flies, long-legged flies, mosquitoes and marsh flies, all of which could have been collected in the wood. Diptera larvae will feed on almost anything: grasses, beech, dung, rotting leaf litter or equally rotting flesh, flowers, fungi, roots … Flies include a huge range of carnivores, scavengers, herbivores, dung-feeders, parasites, bloodsuckers – and it would take a different book to do them justice. Dick Vane-Wright's crane flies have had to stand in for them all. I visualise our woodland habitat crammed with tiny beings in pursuit of dozens of different trades. This is not like a human city, because some of the trades are unknown on our scale. There is no human equivalent of parasitoids; at least, I very much hope not.

I should be more accurate. At insect size the wood is a concatenation of habitats. A nook or a cranny is a habitat, as is the space between two crumpled bramble leaves. Charles Hussey locates a special site I could easily have passed by. Little pools of water collect where beech branches have rotted or broken off, or in holes in stumps. They look dark and unwholesome, as if they could furnish an ingredient for a witches' brew. I should have known: they also provide a special habitat. Charles is a microscopist, and he collects little phials of somewhat murky stuff to take home to examine more closely. He reports that there are creatures I could never have imagined living in these miniature lagoons. A curious pallid, elongate little thing with six legs is a larva of an uncommon beetle, specialised for this location, called *Prionocyphon serricornis*. A few mosquito larvae (*Anopheles*) are less of a surprise, as I am used to seeing them twitching about in wet, stagnant corners at home. I really did not expect to see a crustacean, but there proved to be plenty of minute copepods (*Bryocamptus minutus*). I usually think of such planktonic crustaceans as marine, and they are among the most abundant creatures in the sea, but there are plenty of freshwater species, and now here is a tiny, segmented, hairy-limbed swimming animal specialised for life in tree holes.

Charles finds one example of his own favourite group of miniature organisms: a rotifer. We are now at the edge of visibility, which I set as the boundary for the Grim's Dyke Wood project. And there are many smaller creatures in this living soup. Something resembling a microscopic tulip is a protozoan, probably *Vorticella*, composed of just one cell. Other protists whizz by under the microscope like tiny self-propelled machines, too fast to identify. All these organisms feed on others still smaller, and far too minute to be readily observed beneath Charles's binocular microscope. Many of these are bacteria – another whole universe of diversity – and much smaller again are viruses. I do not doubt that living and reproducing in the tree pools in my wood are species never named by science. An infinite regression of ever-

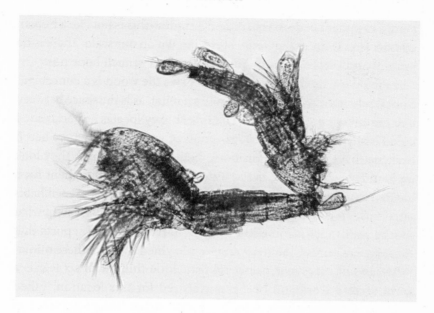

Copepod crustacean *Bryocamptus*, photographed by Charles Hussey.

smaller life forms is revealed. When it rains, millions of tiny things revel briefly in watery profusion in the transitory streams that run down our beech trunks. The soil always heaves with them, and could not exist without them. If anything underlies the real world, these invisible ciphers write the language of complexity, biological motes that interact with, and entrain, everything more complex. No kind of apology is adequate for relegating the 'fundamental particles' of biology to little more than a footnote.

New growth

The woodland is stirring. Rosettes of bluebell leaves have fully erupted, and are so dense in a few parts of the wood that all the ground is richly green. The Dell in the middle of Grim's Dyke is quite the meadow: the special grass that loves beech woods – wood melick – is so lush and bright yellow-green that it is hard to believe its sward will be gone by late summer. Arum leaves are prolific in the shadiest places, precocious wood spurge and celandines are already in flower, and the female flowers of wych elm hang in papery cascades over the path. Wood pigeons perch up there somewhere, calling repeatedly, 'Take *two* pills, David! Take *two* pills, David!' Beech branches are still completely bare, and pearly light from above filters through them, making jagged graffiti against the bright sky. Soon enough, canopy leaves will open in their turn. The wheel of the seasons turns over and over again. It should all feel timeless, but I now understand that nothing in the wood is untouched by history. This ancient ground has been inextricably intertwined with human employment, and economic imperatives have shaped it as much as arboriculture or squirrels. Even the air carries its own subtle influences from far away. If climate change accelerated, the long reign of the beech tree could be over. Whether I like it or not, our small patch of woodland is a tiny bit of an integrated world, and progressively so since the time of the de Greys. I shudder at the thought of our definitive Chiltern Hills beech trees retreating to moist redoubts, as *The New Sylva* predicts, but I have to accept that nothing lasts forever, not even ancient woodland.

Writing on the English countryside is often suffused with another sense of loss. Edward Thomas's rapt descriptions in *The South Country* are tinged with sadness, a feeling of the world gone askew, with the ancient English landscape offering a cure for an intolerable, and usually urban, malaise. H.J. Massingham's *Chiltern Country* is an eloquent rant against the encroachment of bourgeois values at the cost of the honest skills of wise craftsmen in humble cottages – the 'real' people. Massingham mourns the passing of the way 'the heavy bearded hills used to look when the works of man were in harmony with them, when Man and Nature made *one whole*'. The modern world is somehow out of kilter compared with the way things were – when? In medieval times, or in the eighteenth century? It isn't clear. I beg leave to doubt whether those craftsmen who turned out chair legs by the thousand for a pittance, or the wives who went funny-eyed tatting lace to get food on the table would have seen matters in the same way. Massingham wonderfully – but madly – suggests that long-term Chiltern natives may have descended directly from the original, pre-Saxon inhabitants, supposing an unbroken lineage that modern genomics would be capable of detecting in a moment were it true. He was deeply in love with the idea of continuity. An underlying emotion in these ruralist writers is allied to one conveyed by A.E. Housman's famous lines 'That is the land of lost content/I see it shining plain.' Plain it may shine, but the land is a fantasy of the past remembered, reworked and rebuilt into a great poem of regret. I sometimes wonder whether Edward Thomas described, or reconstructed, his walks.

Massingham's landscape is refracted through the eyes of a posh boy in thrall to a holistic fantasy. Some contemporary nature writing is rich in the details of the author sympathising in some fuzzy way with the totality of nature and the interconnectedness of things, but engagement with the nitty-gritty details of living animals and plants is not on the literary agenda. I prefer the eloquence of detail. I believe that all organisms are as interesting as human beings, and certainly no less important than the observer.

A small group of us stand in a circle in the middle of the big clearing. A mass of brambles has been removed, pulled out by the roots, to make a patch of open ground sufficient for planting a tree. I had to purchase a handsomely proportioned young English oak, because an acorn from our own tree would have needed too much time to grow to the right size. We are planting a new oak in memory of a friend who died of cancer. Her family is gathered round, and now the younger members attack the ground with mattock and spade. The clay-with-flints puts up its usual fight. Large knobbly stones heave reluctantly out of the gluey embrace of the orange clay. The spade strikes sparks off a dark broken flint, as it would have done for a Neolithic hunter sharpening his axe-head in the same place millennia before. A liver-coloured pebble reminds us of a time even before the wood, when a frigid climate gripped the whole of Europe. At last, a goodly hole has been excavated. The tree is placed in position, and four stout stakes harvested from our coppiced hazel trees are securely dug in alongside. After the hole has been backfilled and firmed down, the new planting is watered in, and fenced off by attaching a cage of chicken wire to the stakes – it would be too bad if deer damaged the new growth. The tree is in the best place possible, with plenty of blue sky above, and nothing to shade its upward progress. In a century it should be a fine specimen, and in five hundred years it will still be in its prime. But right now it will be a month before the tight oak buds unfurl their fresh leaves. A few words are spoken to formalise the occasion, and everybody is quiet for a minute or two. Birdsong and a faint creaking of boughs supply the only commentary. This, the latest of so many intersections of Grim's Dyke Wood with human history, is the first one marked by a plaque.

Cabinet of Curiosity

Philip Koomen has delivered our collector's cabinet from Wheelers Barn. It has taken him a month to fashion our cherry timber into a neat case with four drawers, every dovetail joint crafted by hand. The cabinet and the stand on which it rests have been made with clean lines. We discussed handles at some length before rejecting the idea of turned knobs in favour of a kind of wave that grows out of each of the drawers, which are of different sizes to accommodate the finds. These do not make up a systematic collection of the kind I worked with all my life at the Natural History Museum in London. It won't be catalogued by genus and species. It won't be arranged in a scientific way. It will be more like the box of small wonders picked up by a child on the beach, the first harvest of curiosity. Nothing in the collection has any monetary value. Each item is a memento of a month in the Grim's Dyke Wood project. Whenever I look at whatever-it-is the moment of its discovery will be recalled; perhaps it should be described as a recollection, rather than a collection.

The drawers have internal divisions to separate the memories one from another. The largest object is a flint that fortuitously resembles an ox – the 'rother' of Rotherfield Greys, to which the wood was attached for so many centuries. The smallest items are cherry-stones nibbled by a wood mouse and left in a neat store by a tree trunk. Here is that empty thrush egg, blue as the sky on the best day we ever enjoyed. Now come the perfectly round 'chalk eggs' – although there is nothing ornithological about them – that sometimes contain a white paste Dr Robert Plot recognised more than three hundred years ago; add a tile that Lonny made from the clay in our clay-with-flints. Dried truffles from the roots of a beech tree; a spider's lair; a gall from our oak tree; a polished section through a holly trunk betraying the secrets of its age in concentric circles; red sandstones from far away and long before the wood was born; knapped flint; mosses in packets – every object has its place. The last item collected is treated with special care: the dormouse nest found a week ago. Into the cabinet goes the small, leather-bound notebook in which I have recorded the seasons passing under the trees: joyful days, moments of discovery, birdsong, or tree-felling. The story of Grim's Dyke Wood goes into a cabinet made from cherry wood nourished in exactly the same place deep within the Chiltern Hills, in our own small and ancient corner of England. The drawers slide closed effortlessly, crafted to secure the memories, feelings and discoveries of a year in the life of our wood. Curiosity is satisfied. For a while.

Acknowledgements

My wife Jackie has been indispensable to the Grim's Dyke Wood project. Not only did she spot the wood for sale in the first place, but she enjoyed all the adventures that followed. She undertook virtually all of the historical research. She took many of the photographs used to illustrate what happened over the year. It would be no exaggeration to say that without Jackie there would have been no book. I thank Leo Fortey for wise computer advice, and for drawing the location map.

Many people helped with identifications of animals and plants belonging to groups in which I have no expertise. I should mention Andrew Padmore first, because his tireless efforts to compile a catalogue of our moth species entailed more on-site visits than anybody else. His wife Clare spent several convivial late evenings with us. Andrew's photographs adorn this book, and I wish there had been room for more. Peter Creed generously guided me through mosses and liverworts, which I could not have attempted myself. Pat Wolseley kindly identified the lichens I collected. Her colleagues at the Natural History Museum identified beetles in variety, and I particularly thank Max Barclay and Michael Geiser for their unrivalled knowledge, ably aided by young Jordan Rainey, coleopterist-in-waiting. Sally-Anne Spence brought her Minibeast Mayhem crew (including Jordan) to the wood, when a good time was had by all, thanks to Sally-Anne's irrepressible enthusiasm – and small mammals duly appeared for inspection. It was, however, Laura Henderson who discovered evidence of dormice on another occasion. John Tweddle added woodlice, centipedes and millipedes to the list. My old friend Dick Vane-Wright allowed the Diptera to make a respectable contribution in the leggy form of crane flies, collected during several visits. On two occasions Andrew Polaszek collected tiny insects by sweeping with a special net, garnering tiny parasitic wasps in particular; some of these were passed on to Charles Godfray for identification. Nobody could have done the work on these insects with more authority. Peter Chandler, Roger Booth, Duncan Sivell and Gavin Broad made more insect additions, while Rony Huys identified a crustacean. Toby Abrehart added the molluscs. Many species of spiders were identified by Lawrence Bee during his three visits, and a few additional ones were added by Paul Selden; I am indebted to them for sharing their experience with me.

The canopy visit and its cherry-picker were arranged by Neil Melleney, himself no mean naturalist. The visit of Natural History Museum scientists on the same day was kindly organised by Daniel Whitmore. Everyone got a turn in the treetops. Claire Andrews arrived in the wood to record bat squeaks, and kindly went through the recordings with the author afterwards to reveal a surprising variety. Although most of the fungi are my own determinations, Alick Henrici helped me with one or two species that had me stumped. Hans-Josef Schroers confirmed two difficult, tiny fungi. Finally, my long-time fellow commuter Charles Hussey collected small organisms from damp hollows at the foot of beech trees to discover yet another ecology of which I had been unaware.

For sharing their memories of Chiltern woods in former times I thank especially David Rose and John Hill. The National Trust staff at Greys Court generously allowed us to examine their old estate maps. Richard Ovey of Hernes – whose family have lived in the area for generations – kindly laid out for us a huge and detailed old map of Lambridge Wood and its surroundings; Richard's wife Gillian also introduced us to Susan Fulford-Dobson, the last member of the Stapleton family living around Henley, the family that owned Greys Court for several centuries. Susan was kind enough to let us use a manuscript history of her family compiled by her father. Rowena Emmett provided useful information concerning Cecil Roberts. The story of our wood is bound up with that of Henley-on-Thames, and much useful material was located in the River and Rowing Museum. The history of the town and its buildings were explained to us in scholarly detail by Ruth Gibson, who knows every old oak beam individually. The Henley-on-Thames Archaeological and Historical Group is thanked for their help. Paul Clayden kindly lent several items from his library – and for a long time. The staff at the County Record Office in Oxford were efficient providers of further information of historical interest. The Museum of Rural Life at the University of Reading kindly allowed access to several rare publications concerning agricultural history. The local studies collection at Henley Library provided additional information. The High Wycombe Museum is acknowledged for its displays about the 'age of the bodgers'. Joanna Cary, Hayden Jones, and Sir James and Monica Barlow each contributed their own stories relating to the wood and its environs. The enthusiastic volunteers of the Chiltern Society restored our footpaths and put up new waymarks.

To help us unscramble still earlier history Jill Eyers, ever resourceful, brought in another team of volunteers to dig out and investigate Grim's Dyke. They contributed to a particularly important episode in our story. Paul Henderson dug holes in several parts of the wood to explore the geology. Jan Zalasiewicz kindly arranged for the cutting and sectioning of interesting pebbles, which helped to elucidate the Ice Age history of the wood, and for advice on their interpretation I am particularly indebted to Professor Phil Gibbard. To illustrate another use of the clay-with-flints Lonny van Ryswyck and Nadine Sterk at Atelier NL in Eindhoven, the Netherlands, successfully experimented in producing tiles cooked from the clays in the wood. Flints were melted down to produce glass in the same studio.

Processing trees for firewood and timber was part of the Grim's Dyke Wood project. John Moorby kindly took down a beech tree which was split for firewood. Martin Drew felled specimen cherry trees, planked them, and brought the planks back to the wood to season. Some of this planked wood was diverted to Philip Koomen to make our cherry-wood cabinet, as described in the book. Andrew Hawkins generously demonstrated how to make charcoal at a small scale from 'trimmings' from beech trees (and very good charcoal too). Alistair Phillips showed us the finer points of turning wood into bowls – which we still use and treasure.

Lucius Cary made us another one just for fun. John Morris, director of the Chiltern Woodlands Project, advised us on the principles of woodland management, and some of that advice we will attempt to follow. Photographs were taken by Birgir Bohm DFF through the seasons to capture the changes in the wood from one particular viewpoint. I thank him for his patient contribution. Robert Francis supplied several more excellent photographs, which have improved the illustrations. Our 'woody neighbours' bought other parts of the former Lambridge Wood, and they have all been a pleasure to get to know. In particular, Nina Krauzewicz made beautiful drawings of a few of our plants.

Heather Godwin read the first draft of the book, as she has so often before. Her suggestions for improving the clarity or punch of the text are always wise, and usually followed. Robert Lacey read and improved the manuscript with exemplary care. Finally, Arabella Pike has been my loyal editor for many years. I thank her for her patience, and hope she finds this book worthy of its predecessors.

Atelier NL, Eindhoven

Atelier NL, a Dutch design studio, was asked to visualise the material transformation of one square metre of earth from Grim's Dyke Wood. The native earth and stone yielded a unique array of pigments, ceramics and glass. The excavation process clarified the history of the land, which has never been farmed. The flint-studded clay was stiff and difficult to penetrate. The flint sparked as it was crushed and ground into a fine powder. When exposed to high temperatures, the flint powder melted into white, green and bluish glass. The tough earth, kneaded, cleaned and dried, was exposed to a gradation of extreme temperatures. A bold array of pigments and ceramic tiles was created as a result, ranging from warm creamy brown to rich peat. In one small patch of ground, all of the resources were found to make a variety of paints, pigments, ceramic tiles and glass. Atelier NL's work at Grim's Dyke Wood reflects humanity's long relationship with the transformation of land. Through digging, sorting, grinding and heating, the geological history of land merges with story of human creation. (Text: Liz Holland)

Notes

1: April

1. Clare Leighton. *Four Hedges: A Gardener's Chronicle*. Victor Gollancz. 1935.
2. Edward Thomas. *The South Country*. 1909.
3. The Chalk is often capitalised because the name refers to a specific rock formation of Cretaceous age deposited 100–65 million years ago. Informally, 'chalk' can be applied to describe an unusually pure, white and usually soft limestone.
4. The highest point on the Chiltern Hills is at Haddington Hill, Wendover, twenty-five miles away in Buckinghamshire, at 876 feet (267 metres). Most of the wood is at about 360 feet.
5. A.G. Street. *Hedge Trimmings*. Faber & Faber. 1933.
6. Henry David Thoreau. *Walking*. 1862.
7. Robert Gibbings. *Sweet Thames Run Softly*. Dent. 1940.
8. *The Victoria History of the Counties of England: A History of the County of Oxford*. Vol. XVI. 2011.
9. ORO Map 2101M, apportionment 2101A, 1842.
10. Barlow archive, courtesy Monica Barlow.
11. John Morris. *The Cultural Heritage of the Chiltern Woods*. Chilterns Woodland Project. 2009.

2: May

1. It has another common name, 'chicken-of-the-woods', and many people enjoy eating it in the young state. However, some people react badly to this species, so no recipe for it will be given here.
2. We have also found a related species, *Lycogala terrestris*, in the wood.
3. The habits of the celandine were described in a rather pedestrian, if botanically appropriate, verse by Wordsworth:
   There is a Flower, the Lesser Celandine,
   That shrinks, like many more, from cold and rain;

And, the first moment that the sun may shine,
  Bright as the sun himself, 'tis out again!

4. Robert Calder. *Beware the British Serpent: The Role of Writers in British Propaganda in the United States 1939–1945.* McGill-Queen's University Press. 2005.

5. H.J. Massingham. *Chiltern Country.* The Face of Britain. Batsford. 1940.

6. J.S. Mill. 'Walking Tour of Berkshire, Buckinghamshire, Oxfordshire and Surrey, 3–15 July'. *Collected Works of John Stuart Mill* (ed. J.M. Robson). Routledge & Kegan Paul. London. 1988.

7. George Grote. *A History of Greece: From the Earliest Period to the Close of the Generation Contemporary with Alexander the Great.* 1846–56.

8. Harriet Grote. *The Personal Life of George Grote, Compiled from Family Documents, Private Memoranda, and Original Letters to and from Various Friends.* John Murray. 1873.

9. Richard Mabey. *Home Country.* Random Century. 1990.

10. Geological Survey of England and Wales. 1:50,000 Geological Map series. Map 254. Henley-on-Thames.

11. There is nothing new about my discovery. Mr Llewellyn Treacher refers to these pebbles in a geological supplement appended to Emily Climenson's guide to Henley-on-Thames of 1896. He says that 'there are several sections which well show the character of this gravel in Lambridge Wood'. They are not there now.

12. The modern name for these rocks is the Kidderminster Formation.

13. P.L. Gibbard. The history of the great north-west European rivers during the past three million years. *Philosophical Transactions of the Royal Society of London* B318. 559–602. 1988.

14. Oliver Rackham. *Trees and Woodland in the British Landscape.* Revised edition. Dent. 1990.

15. Alexander Allardyce (ed.). *Letters from and to Charles Kirkpatrick Sharpe, Esq.* William Blackwood & Sons. 1888.

16. The same newspaper report features winners that were familiar in 'posh' Henley rural society, the Oveys, the Freemans and the Camoys among them. It seems that 'fiddle faddling' was considered rather a fashionable occupation despite Mr Sharpe's patronising diagnosis.

17. It was formally named in *Geraniaceae* Vol. 3. Robert Sweet. 1826.

3: June

1. Kalm's account of his visit to England on his way to America in 1748. Trans. J. Lucas. Macmillan & Co. New York. 1893.

2. Gabriel Hemery and Sarah Simblet. *The New Sylva: A Discourse of Forest and Orchard Trees for the Twenty-First Century.* Bloomsbury. 2014.

3. I should mention that the same informal name is used for some very different exotic orchids.

4. M. Forty and T.C.G. Rich (eds). *The Botanist: The Botanical Diary of Eleanor Vachell (1879–1948)*. National Museum of Wales. Cardiff. 2006.
5. V.S. Summerhayes. *Wild Orchids of Britain*. New Naturalist. Collins. 1951.
6. M. Jannink and T. Rich. Ghost Orchid rediscovered in Britain after 23 years. *Journal of the Hardy Orchid Society*. Vol. 7. 14–15. 2010.
7. C.M. Cheffings and L. Farrell (eds). *The Vascular Plant Red Data List for Great Britain*. Joint Nature Conservation Committee. 2005.
8. M.I. Bidartondo and T.D. Bruns. Extreme specificity in epiparasitic Monotropoidiae (Ericaceae): widespread phylogenetic and geographical structure. *Molecular Ecology*. 2001.
9. I must give the technical term for this esoteric botanical ploy: mycoheterotrophy. Extraordinary as it is, it has evolved in several more plant families, although most commonly in the Orchidacea.
10. Much later, in some cases. I was still awaiting identifications as this book went to press.

4: July

1. L.W. Hepple and A.M. Doggett in *The Chilterns* (1994) refer to an Anglo-Saxon boundary charter where a section of the ditch is called 'Ealden Wege' (Old Way).
2. R. Bradley. The south Oxfordshire Grim's Ditch and its significance. *Oxoniensis*. Vol. 33. 1–12. 1968.
3. Angela Perkins. *The Phyllis Court Story: From Fourteenth-Century Manor to Twentieth-Century Club*. Phyllis Court Members Club. 1983.
4. This result from *Oxford Ancestors* was not a surprise. My father was from Worcestershire near the Welsh border, and always claimed he was the only Fortey ever to have left the village. Some of these were ditch-diggers.
5. I should note that far older, Palaeolithic flint workings are well-known nearby, as at Highlands Farm just outside Henley. These originated from an ancient human occupation of Britain long before the last Ice Age with which my story begins.
6. J.G. Evans. *Land Snails in Archaeology*. Academic Press. 1972; *The Environment of Early Man in the British Isles*. Paul Elek. 1973.
7. Fred Hageneder. *Yew*. Reaktion Books. 2013
8. Ian D. Rotherham. The ecology and economics of Medieval deer parks. *Landscape Archaeology and Ecology*. Vol. 6. 86–102. 2007.
9. Elizabeth Craig. *Court Favourites: Recipes from Royal Kitchens*. André Deutsch. 1953.

5: August

1. The poetic reader may prefer Sir Sacheverell Sitwell's 1972 description from *Fungi: A Look at Coloured Figures of English Fungi or Mushrooms by James Sowerby F.L.S.* London. 1797:

   A fungus that is egg-shaped in its beginnings,
   but like a goose's egg most typically
   Foetid from the thick layer of green jelly
   Which attracts the flies to feed upon the slime
   and spread about the spores;
   When ripe, the white skin or 'veil' torn
   in monstrous mock-circumcision
   And erect to full height in a matter of a few hours,
   horned god of the coven.

2. The term 'foray' has been used for more than a century, and I prefer it to 'forage', which implies that the only interest is in finding mushrooms that are edible.

3. Roger Kendal. A Romano-British building at Bix, Oxon. Notes on an excavation carried out 1955–1956. *Henley-on-Thames Historical and Archaeological Society*. Vol. 21. 2006.

4. J.E. Eyers (ed.). *Romans in the Hambleden Valley: Yewden Roman Villa.* Chiltern Archaeology Monograph 1. 2011.

5. S.S. Frere. 'The End of Towns in Roman Britain'. In J.S. Wacher (ed.). *The Civitas Capitals of Roman Britain*. Leicester University Press. 1966.

6. H. Cam. *Liberties and Communities in Medieval England.* Cambridge University Press. 1944.

7. L.W. Hepple and A.M. Doggett in *The Chilterns* (1994) convincingly suggest that Hundreds preceded parishes, rather than the former being grouped from previous ecclesiastical units.

8. A notice in the church of St Botolph in Swyncombe tells us that bones of boar have been recovered from the site of the manor.

9. Some scholarly opinion disputes that Yggdrasil was an ash tree, although it is apparently so described in the *Poetic Edda*.

10. By another delightful coincidence: the account of clay industries in Oxfordshire. James Bond et al. *Oxfordshire Brickmakers*. Oxfordshire Museums Service Publication No. 14. 1980.

6: September

1. Richard Mabey. *Flora Britannica*. Sinclair Stevenson. 1996.

2. Michael Macleod. *Land of the Rother Beast: An Oxfordshire Chronicle.* Skye Publications. 2000.

3. A full discussion of this issue is in Simon Townley. *Henley-on-Thames: Town, Trade and River*. Phillimore & Co. 2009.

4. London had an estimated population of up to 100,000 people by 1300. In Europe only Paris was more populous.

5. This magnificent structure is well described by W. B. Logan in *Oak: The Frame of Civilization*. Norton. 2005.

6. The original version was penned by the famous actor David Garrick in 1759.

7. John Steane. Stonor. A lost park and a garden found. *Oxoniensis*. Vol. 59. 1994.

8. At the same time as Henley-on Thames was beginning to prosper, the Charter of the Forest (1217) guaranteed rights such as pannage and estovers for freemen in the royal forests. In private estates such rights were often tradable commodities.

9. This was resolved by simple microscopy. The ascomycetes have their spores carried in sac-like asci, while basidiomycetes bear their spores on the tips of cylindrical basidia.

10. This does not include the flowering plants, which are generally better-known. Fungi, however, are less comprehensively known even than insects.

7: October

1. Notified under the Wildlife and Countryside Act of 1981. Originally notified in 1952. So the wood has been protected from development for more than sixty years.

2. Miles Stapleton. *History of Greys Court*. Unpublished MS in the ownership of Susan Fulford-Dobson.

3. Sally Varlow. Sir Francis Knollys's Latin Dictionary: new evidence for Katherine Carey. *Institute of Historical Research*. 1–9. Blackwells. Oxford. 2006.

4. Simon Townley. *Henley-on-Thames: Town, Trade and River*. Chapter 5. Phillimore & Co. 2009.

5. This candle-snuff fungus, *Xylaria hypoxylon*, is an ascomycete with a complex life history. The stage I observe is the asexual phase, producing white spores. The sexual stage is all black and less conspicuous.

6. These common names for fungi are something quite new. There are relatively few traditional English names for mushrooms, and the ones employed are from L. Holden. *Recommended English Names for Fungi in the UK*. English Nature. 2006.

7. The cep, porcini, Steinpilz, Karl-johan sopp, penny bun, *Boletus edulis* in French, Italian, German, Norwegian, English, and the Latin name Boletus.

8. Recent work has shown that the honey fungus is a complex of species. The one in the wood seems to be the true *Armillaria mellea*.

9. The name 'wych' is derived from an Old English word referring to the flexibility of the twigs.

10. This disease is often reported as being caused by *Chalara fraxinea*, which is the asexual state of *Hymenoscyphus pseudoalbidus* – the name makes no difference to the lethalness.
11. There are at least four species of orb-weaving spiders, including ones like *Metellina* with more slender bodies than *Aranaeus*.

8: November

1. *Oxford Gazette*. 7 January 1854.
2. Parish register. 23 January 1643. 'This day were buried six soldiers whereof four were slaine with the discharge of a cannon as they marched up Duck [Duke] Street to assault ye town.'
3. Ruth Spalding (ed.). *Diary of Bulstrode Whitelocke 1605–1675*. Oxford University Press. 1990.
4. This stylish item of Roundhead gear can be purchased from specialists in period costumes for re-enacting historic battles.
5. Pepys papers. 'Notes about firewood taken at Henley'. Bodleian Library. Ms Rawlinson A 171 f222v.
6. Another portion of the same legacy went to the parish of Bladon, north of Oxford.
7. The enzyme laccase 'snips' the complicated three-dimensional lignin structure, allowing further enzymes and oxygenation to do the rest.
8. P.D. Gabbut. Quantitative sampling of the pseudoscorpion *Chthonius ischnocheles* from beech litter. *Journal of Zoology*. Vol. 151. 469–78. 1967. Like mites, pseudoscorpions are arachnids.
9. Nonetheless, it is known that at least forty species of nematodes are confined to living on rotten wood.
10. I am conscious that I have not listed the different species of worms, slugs and beetles (and more) under the same log. That deficiency will be repaired online.
11. Bodleian Library, Oxford. AASHM 1677 (14). A more measured report in the *Philosophical Transactions* for 1683 says that it lasted for six seconds.
12. P.D. Elliman. *Glassmaking in Henley-on-Thames in the Seventeenth Century*. Henley Historical and Archaeological Society. 2–15. 1979.
13. I should note that Ravenhead Glass is a much later manufacturer.

9: December

1. This was already big business in 1667, when a Henley merchant, George Cranfield, had '800 longe velleyes' ready to ship to London. Pat Preece. Wheelwrights. *S. Oxfordshire Archaeology Bulletin*. Vol. 59. 2004.
2. Daniel Defoe. *A Tour Thro' the Whole Island of Great Britain*. Letter 4. Part 3. Berkshire & Buckinghamshire. 1726.
3. To be accurate, I should mention that Henley has its fair share of false

(or blocked) windows to maintain symmetry, but avoid having to pay the Window Tax that was current in the eighteenth century.

4. Richard B. Sheridan. *Sugar and Slaves: An Economic History of the British West Indies 1623–1775*. Canoe Press. 1974.

5. Keith Mason. The world an absentee planter and his slaves made. Sir William Stapleton and his Nevis Sugar Estate, 1722–40. *Bull. J. Rylands Library Liverpool*. Vol. 75. 1993.

6. There were older-established Hellfire Clubs, but the fame of Dashwood's club seems to have eclipsed the others.

7. It was. Archaeologists uncovered the contents of the 'time capsule' early in the twentieth century, and they are now on display in the River and Rowing Museum.

8. J. Thirsk. *The Agrarian History of England and Wales*. Vol. 5. Oxford University Press. 1984.

9. L.W. Hepple and A.M. Doggett in *The Chilterns* emphasise that enclosure happened over three centuries, and continued well into the nineteenth century. In this case, agricultural 'revolution' seems more like enforced evolution.

10. Robert Phillips. *Dissertation concerning the Present State of the High Roads of England, especially of those near London*. 1736.

11. Another large estate adjacent to the River Thames at Whitchurch. Mrs Powys was the wife of Philip Powys.

12. Cecil Roberts. *Gone Afield*. Hodder & Stoughton. 1936.

13. Lord Wyfold. *The Upper Thames Valley: Some Antiquarian Notes*. George Allen & Unwin. 1923.

14. These spores are borne in sac-like asci, which identifies the fungal partner as an ascomycete, and a distant relative of morels.

10: January

1. William Black. *The Strange Adventures of a Phaeton*. Street & Smith. 1872.

2. Improvements in road-building techniques were developed by the Macadam family: John L. Macadam and his son James, 'the colossus of roads'.

3. L.J. Mayes. *The History of Chairmaking in High Wycombe*. Routledge & Kegan Paul. 1960.

4. These were trimmings of side shoots from trees to help straight growth. 'Shrouding faggots' were derived from the crowns of trees after felling.

5. *Jackson's Oxford Journal*. No. 4946. 12 February 1848.

6. J. Morris. *The Cultural Heritage of Chiltern Woods*. Chiltern Woodlands Project. 2009.

7. Cecil Roberts. *Gone Rustic*. Hodder & Stoughton. 1934.

8. Not enough to feed a family. Angela Spencer-Harper. *Dipping into the Wells*. Robert Boyd Publications. 1999.

9. Turned wares were already in use in medieval times, although the Worshipful Company of Turners did not get their Charter until 1604.

10. He served as an officer of the Royal Microscopical Society, which is still going strong today, to which he made generous donations.
11. Sally Strutt. *History of the Culden Faw Estate*. Privately published. 2013.
12. I have an old book written by a friend of my father, Eric Ennion, to help me identify prints. N. Tinbergen and E.A.R. Ennion. *Tracks*. Oxford University Press. 1967.

11: February

1. Peter Creed and Tom Haynes. *A Guide to Finding Mosses in Berkshire, Buckinghamshire and Oxfordshire*. Pisces Publications. 2013.
2. Steam traction engines were heavy. A notice survives on Henley Bridge permitting only one vehicle at a time.
3. I am indebted to Angela Spencer-Harper's book *Dipping into the Wells* for this information.
4. The official history of the company is called, inevitably, *A Brush with Heritage*. Christine Clark. Centre of East Anglian Studies, University of East Anglia. 1997.
5. Francis Sheppard. *Brakspear's Brewery Henley-on-Thames 1779–1979*. Published by the brewers. 1979.

12: March

1. J.H. Van Stone. *The Raw Materials of Commerce*. Vol. 1. Pitman & Sons. 1929.
2. Berkshire, Buckinghamshire and Oxfordshire Wildlife Trust (BBOWT); and a delightful place to see many species typical of chalk countryside.
3. Paul Bright and Pat Morris. *Dormice*. The Mammal Society. 1992.
4. It is quite difficult to find a working barn in the Chiltern Hills. There is one in the Hambleden Valley, and another near Harpsden, and (of course) at Stonor. I await their conversion.
5. Olivia Harrison was justifiably nervous after a violent attack on George by an intruder in 1999, and I am sure that razorwire would not have surrounded Friar Park in other circumstances.
6. As this book went to press the total had surpassed 150.
7. C.E. Hart. *Timber Prices and Costing 1966–67*. Published by the author.
8. An exhibition at the River and Rowing Museum in Henley suggests that Fawley Court may have been the inspiration for Toad Hall.
9. Gabriel Hemery, the author of *The New Sylva*, is its visionary. The charity was set up in 2008 to revive interest in native woodland, to explore old and invent new industrial uses for wood products, to inspire schoolchildren, and to engage artists and scientists in the cause.
10. Manuel Limo. *The Book of Trees: Visualizing Branches of Knowledge*. Princeton Architectural Press. 2013.

Index